AF544175

EUL
VERLAG

Kathrin Bischoff

Business Approaches to Poverty Alleviation

A Classification of Company Initiatives

With a Preface by Prof. Dr. Dr. h. c. Norbert Szyperski, University of Cologne

Bibliografische Information der Deutschen Nationalbibliothek

Die Deutsche Nationalbibliothek verzeichnet diese Publikation in der Deutschen Nationalbibliografie; detaillierte bibliografische Daten sind im Internet über <http://dnb.d-nb.de> abrufbar.

ISBN 978-3-8441-0393-9
1. Auflage Dezember 2014

JOSEF EUL VERLAG GmbH
Brandsberg 6
53797 Lohmar
Tel.: 0 22 05 / 90 10 6-6
Fax: 0 22 05 / 90 10 6-88
E-Mail: info@eul-verlag.de
http://www.eul-verlag.de

Bei der Herstellung unserer Bücher möchten wir die Umwelt schonen. Dieses Buch ist daher auf säurefreiem, 100% chlorfrei gebleichtem, alterungsbeständigem Papier nach DIN 6738 gedruckt.

Preface

Due to its prevalence and persistence, poverty remains a topic of utmost relevance. Research in the field of poverty alleviation tends to be widespread but partially inconclusive, in particular with regard to the role of the private sector in combating the issue of poverty.

The study of Anne Kathrin Bischoff therefore addresses an important research theme and presents a comprehensive and innovative approach towards the analysis of business initiatives towards poverty alleviation. The author uses a highly structured, analytical and coherent approach in her work. The analysis is well founded on fundamental research questions and high quality literature. The theoretical and practical methodology and especially the structured literature review attest the pronounced research skills of the author. The developed classification framework is innovative and persuasive. All findings are well presented in a structured, illustrative and comprehensive manner.

This work provides an excellent decision guiding tool for companies that want to improve their poverty alleviation strategy. Overall, the study of Anne Kathrin Bischoff is an outstanding contribution to the existing research on business approaches to poverty alleviation.

Cologne, December 2013 — Prof. Dr. Dr. h. c. Norbert Szyperski

Acknowledgments

This Master thesis has been accepted by the Rotterdam School of Management of the Erasmus University Rotterdam in December 2011.

There are several people I would like to thank for their support during my thesis writing process:

First of all, I want to thank Andrea da Rosa for her excessive feedback and assistance on a day-to-day basis. I would of course also like to express my gratitude to my coach, Prof. Dr. Rob van Tulder, Erasmus University Rotterdam, and my co-reader, Prof. Dr. Slawomir Magala, Erasmus University Rotterdam, for all the time and effort spent on guiding me through this process.

Personally, I would like to thank my parents, Dr. Margot Eul and Prof. Dr. Johannes Georg Bischoff, for their continuous support and for enabling me all the possibilities I have had this far. I would not have been able to get where I am now without this encouragement and support. I would also like to show my appreciation for the constant encouragement from my friends, especially Shara Darr and Stephanie Kuhl. I would like to thank Alexander Schabel for always being there.

Ultimately, I would like to thank Prof. Dr. Dr. h. c. Norbert Szyperski, University of Cologne, for the acceptance of my thesis within the Interscience scientific series as well as the employees of the Josef Eul Verlag for the great support during the publication of my thesis.

I am very grateful to all of the above mentioned persons and I would not have been able to write my thesis without you.

Cologne, December 2013 Anne Kathrin Bischoff

Table of Contents

List of Figures

List of Tables

List of Boxes

List of Abbreviations

BOP	Bottom of the Pyramid
CP	Corporate Philanthropy
CSO	Civil Society Organizations
CSP	Cross-Sector Partnerships
CSR	Corporate Social Responsibility
FDI	Foreign Direct Investment
MDG	Millennium Development Goal
MNC	Multinational Corporation
NGO	Non-Governmental Organizations
PPP	Public-Private Partnerships
SE	Social Entrepreneurship
SME	Small and Medium Enterprises

1 Introduction

Poverty remains a prevailing and urgent issue around the globe. According to economic poverty measures by the World Bank, approximately 1.4 billion people live on less than $1.25 a day, which represents about 27% of the world's population (World Bank, 2008). More recently, multidimensional poverty estimates reveal even more striking results. A respective example is the Multidimensional Poverty Index, which measures poverty based on ten indicators within health, education and living standards dimensions. Accordingly, roughly 1.7 billion people are classified as living in multidimensional poverty, which implies that "...at least 33 percent of the indicators reflect acute deprivation in health, education and standards of living" (UNDP website, 2011).

Despite substantial initiatives undertaken in the last decades by governments and Non-Profit Organizations (NGOs), high poverty rates prevail. A fruitful explanation has thus been the inability of government and the non-profit sector to address the issue of poverty in isolation (Hahn, 2009). Hence, two promising shifts in the field of development and poverty alleviation have recently taken place. Hereby, the first change refers to the greater employment of market-based approaches and the increased involvement of the private sector, in particular Multi-National Corporations (MNCs), in development activities (Newell and Frynas, 2007). The second emerging trend in turn encompasses the enhanced focus on cross-sector collaboration between societal actors – being the state, the civil society and the market – in order to jointly address societal issues, such as poverty (Chakraborty et al., 2004).

As a result, the increased involvement of, as well as the collaboration with the private sector have triggered the hope of ultimately increasing the effectiveness of poverty alleviation initiatives. However, at present, widespread ambiguity remains on how the private sector can best be incorporated into the field of development. Accordingly, the overall contribution of the business sector, in general, and MNCs, in particular, as well as the potential poverty-alleviating effect of such business strategies and cross-sector partnerships remains controversial and ill-understood. (Campbell, 2005; Prieto-Carrón et al, 2006; Kolk and van Tulder, 2006)

1.1 Research Questions and Aim

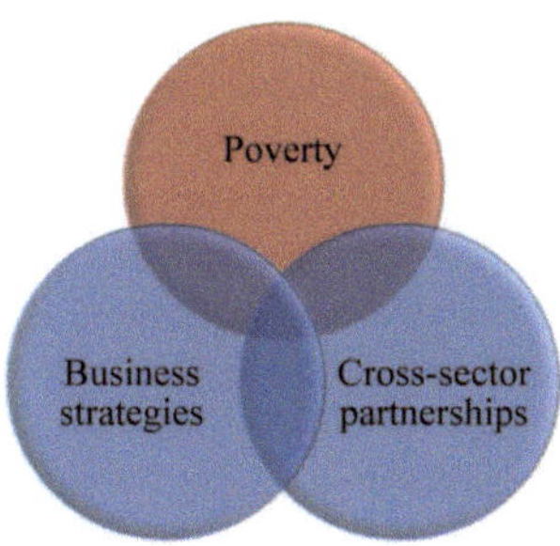

Figure 1-1: Conceptualization of Themes Addressed

As a result, this thesis analyzes the following three themes separately and jointly, namely *poverty*, *business strategies* for combating poverty and *Cross-Sector Partnerships* (CSPs) towards tackling poverty. Hereby, the focus is placed on the business strategies and cross-sector partnerships which directly aim at addressing poverty. A visual conception of the target areas and their overlaps is displayed in red in Figure 1-1, being poverty; business strategies *and* poverty; CSPs *and* poverty; as well as business strategies *and* CSPs *and* poverty.

Overall, this thesis aims at shedding light on the research area of business and poverty by investigating the potential and actual contributions of the business sector, in general, and MNCs, in particular, on poverty alleviation.

Accordingly, the following six research questions are addressed:

- *What are the roles and potential contributions of the business sector, in general, and MNCs, in particular, towards combating poverty?*
- *In how far are the examined business strategies on poverty alleviation – being Business-as-Usual; Corporate Philanthropy; Microfinance; Bottom of the Pyramid and Social Entrepreneurship – and their variations effective in addressing poverty?*
- *In how far are Cross-Sector Partnerships effective in addressing poverty?*
- *In how far can the prevailing poverty alleviation initiatives be combined and in how far would such a joint approach increase the effectiveness of the undertakings?*
- *Which poverty-related initiatives are actually undertaken by Fortune's largest 100 MNCs?*
- *In how far do the Poverty Alleviation Portfolio Approaches by the largest 100 MNCs reach up to the theoretical potential of the private sector in effectively addressing poverty?*

In order to answer these questions, theoretical and practical research is conducted. Hereby, the theoretical structured literature review examines the first four research questions by providing an overview of academic views on the topic whereas the practical research, being conducted on Fortune's largest 100 MNCs, investigates the final two research questions.

1.2 Analysis Approach

Overall, regarding the analysis approach undertaken in this thesis, several remarks have to be made. Concerning the unit of analysis, it should be highlighted that poverty is an issue which generally affects all societal actors. According to the societal triangle, as displayed in Figure 1-2, society can be sub-divided into three different spheres which each contain several actors (van Tulder and van der Zwart, 2006). The first sphere in this regard is the state whose primarily responsibility lies in establishing a solid legal framework and an equitable and inclusive social environment (van Tulder and van der Zwart, 2006). In the context of poverty, actors of vital importance are national and local governments, international organizations as well as bilateral or multilateral donors. For simplicity, these actors are simply referred to under the umbrella term 'government' in succeeding discussions. Next, the second sphere of the societal triangle relates to the civil society domain, whose primary duty can be regarded as fostering relationships between distinct groups as well as supporting marginalized groups which are disregarded by the two other sectors (van Tulder and van der Zwart, 2006). Main actors of interest with regard to poverty are Nonprofit or Nongovernmental Organizations – being jointly denoted as 'NGOs' – as well as the poor themselves and their communities. Lastly, the third sphere corresponds to the market, whose main role is to transform inputs into outputs in order to boost the economy of a country in an inclusive and sustainable manner (van Tulder and van der Zwart, 2006). Essential actors in this respect are Multinational Corporations (MNCs) which operate in developing countries or in poor communities as well as local business, Small-and-Medium Enterprises (SMEs) and entrepreneurs which originate from poor regions.

As displayed in red in Figure 1-2, this thesis analyzes the issue of poverty alleviation by focusing on the *market sphere* of the societal triangle, while still including the interactions of companies with state and civil society actors in discussions. Hereby, emphasis is placed particularly on the potential contributions and approaches of *MNCs*, in contrast to small, local firms, towards tackling poverty. This focus is employed due to the increased role which has been attributed recently to the business sector, in general, and powerful MNCs, in particular, within the fields of development and poverty (Hahn, 2009; Kolk and van Tulder, 2010).

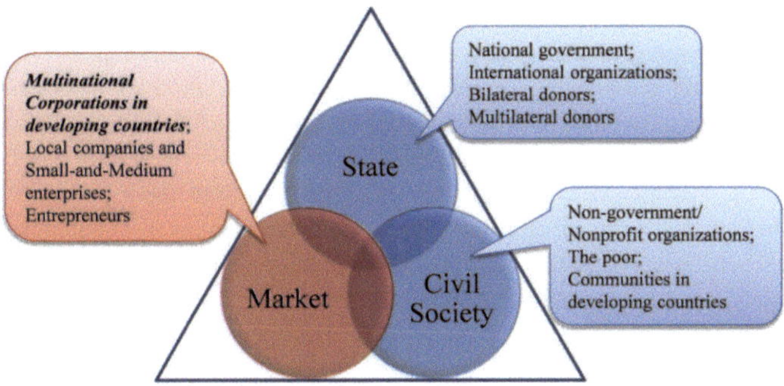

Figure 1-2: Actors in the Societal Triangle with Regard to Poverty (based on van Tulder and van der Zwart, 2006)

In order to investigate the potential contribution of MNCs towards poverty reduction, a discussion of the role of companies in solving societal problems, such as poverty, is provided. This analysis aims to answer *whether* firms should engage into poverty eliminating business activities. Afterwards, it is examined *how* companies should do so. Accordingly, a selection of five business strategies – Business-as-Usual, Corporate Philanthropy (CP), Microfinance, Bottom of the Pyramid (BOP) and Social Entrepreneurship (SE) – as well as the concept of Cross-Sector Partnerships (CSPs) are presented as possible options. These approaches are chosen as subjects of analysis on basis of a structured literature search and the ultimately selected keywords and articles. Hence, for feasibility reasons it is required that the examined initiatives have been extensively discussed among academics. Moreover, a further selection criterion is the applicability of the business approaches to MNCs, as this is the focus of the study. An examination of these approaches in turn leads to a classification of approaches in terms of their effectiveness for poverty alleviation. Thus, it is noteworthy to highlight that the selected business initiatives on poverty reduction are analyzed according to their *impact on the poor and society* in general and not according to the implications of such engagement for business performance.

On the whole, this thesis covers an intensive range of complex, inter-disciplinary themes. As a result, an in-depth analysis of each single theme discussed cannot be provided. Instead, this thesis aims at employing a *helicopter perspective* on the area of business and poverty alleviation. The goal is to generally investigate present initiatives in order to classify, connect and compare them. Such a joint examination of existing business approaches on poverty alleviation shall consequently present the 'bigger picture' in order to facilitate an overall understanding of the research field.

1.3 Relevance of the Study

The literature review in this thesis reveals that the business sector, in general, and MNCs, in particular, are widely regarded as "...part of the solution to the problem of world poverty" (Prieto-Carron et al., 2006, p. 980). Hence, while consensus on *whether* MNCs should be involved in development and poverty alleviation seems to be established, substantial ambiguity among researchers remains on *how* business can most effectively be engaged in poverty-related activities. To put it into the words of Kolk and van Tulder (2006, p. 790): "While the potential and opportunities of MNCs in relation to poverty alleviation are receiving considerable attention, the exact components for fulfilling such a role, in the context of international efforts to delineate poverty and the contribution of MNCs in this regard have hardly been addressed." In related terms, Prieto-Carron et al. (2006, p. 981) draw the following conclusion: "In sum, we believe that a critical research agenda should conduct an in-depth investigation of what actually contributes 'business' and 'poverty' and how different types of businesses may affect different types of poverty both positively and negatively."

Investigating the contribution that MNCs should have on poverty reduction is in theory and in practice highly relevant for several reasons. First, it can provide a first evaluation of the effectiveness of distinct business approaches – as well as their variations and combinations – for poverty alleviation. Such an evaluation in turn could offer guidance to business managers, who want to become more engaged in the field, on which approaches to pursue. Moreover, such an analysis could also serve as a starting point for further joint classifications or comparisons of initiatives in the field of academia. Besides, by revealing how engaged MNCs are in practice in the field of poverty, both managers and researchers can retrieve a first overview of operations in practice which in turn can serve as a tool for firm comparisons. Additionally, by using this study as a basis for further research, the employed company classification could be utilized to identify neglected fields, business opportunities, challenges or best practices in the field. Consequently, this research should enhance understanding in the field of business and poverty and shall thereby foster the full exploitation of the potential contribution of MNCs regarding poverty alleviation.

1.4 Outline

The thesis consists of four chapters. It is structured as follows:

This first chapter serves as an introduction by providing a general overview of the topic and the aspects discussed. More precisely, the three different main aspects discussed in this thesis are briefly presented, being poverty and its alleviation, business strategies on tackling poverty as well as Cross-Sector Partnerships which aim at addressing poverty. Moreover, the research questions, the analysis approach as well as the relevance and outline of the study are presented.

Next, chapter two comprises the theoretical part of the thesis and discusses the findings of the structured literature review. It is composed of four sections. The first section displays the methodology used in conducting a structured literature review. Criteria for the selection of the articles are presented and an overview of the characteristics of these chosen articles is provided.

The second section clarifies the issue of poverty and reveals common approaches towards its alleviation in order to facilitate understanding of overall topic and its context. It follows an examination of the roles and responsibilities of the societal actors with regard to poverty alleviation. More precisely, the concepts of civil society, governmental and market failure are introduced and the recent shift in development activities towards a greater emphasis on business inclusion is presented. This section ends with an examination of the distinct viewpoints on the potential contributions and limitations of business actors with respect to tackling poverty.

This is followed by section three which zooms in on five predominant business strategies towards addressing poverty. First, a clarification of relevant terms and the employed classification framework are presented. Next, the various strategies and their respective variations are briefly explained, their impact on poverty is highlighted and diverging views on their support and criticism are discussed. The business strategies presented encompass the following: Business-as-Usual; Corporate Philanthropy (CP); Microfinance; Bottom of the Pyramid (BOP) and Social Entrepreneurship (SE). This section ends by providing a classification of these business initiatives in terms of their potential towards long-term poverty alleviation.

Afterward, section four focuses on the role of Cross-Sector Partnerships (CSPs) on combating poverty in addition to the notion of Poverty Alleviation Portfolios. Hereby, the concept of CSPs is explained and the potential contributions and limitations of distinct partnering approaches on poverty are presented. Besides, it also includes an analysis of the limited literature available on the joint combination of approaches. Hence, the role of CSPs within the presented business strategies is discussed and the potential combinations among these five business strategies are investigated. Ultimately, a classification on companies' Poverty Alleviation Portfolio Approaches is provided based on all discussed initiatives. This section closes chapter two via a brief conclusion which aims at summarizing the main findings and answering the four research questions of the theoretical part.

Chapter three in turn entails the practical research of this thesis and consists of three sections. In this regard, section one outlines the research methodology. Primary research is conducted by reviewing the websites and CSR reports of Fortune's largest 100 MNCs on their undertakings in terms of the analyzed business approaches on poverty alleviation. With regard to the methodology, first of all, the research approach and the research method are presented, being followed by the data collection method, the sample design and the conceptual model. Afterwards, the approaches for data collection, data codification and data classification are outlined. Lastly, the data analysis and the exploratory analysis are described and research characteristics, in terms of reliability, replication, validity and confirmability, are discussed.

In turn, the second section presents the research findings. Hereby, first overall observations are outlined and an overview of companies' provisions of general poverty statements is provided. Next, the engagement of the examined firms into each of the presented business strategies, namely Business-as-Usual, CP, Microfinance, BOP and SE, as well as their variations is described. This is followed by an overview of the CSP approaches being undertaken by the companies. Next, the results on the simultaneous combination of the business approaches are revealed. Ultimately, the overall Poverty Alleviation Portfolios per sample firm are classified according to the developed framework in the theoretical part.

Lastly, section three discusses the findings of the practical part. To begin, a general discussion of the findings is presented. Next, possible underlying reasons for differences in companies' Poverty Alleviation Portfolio Approaches are explored. Hereby, the examined factors are company sector, region or country of origin as well as company narrative, which is comprised of the issue framing and the issue prioritization approach on poverty. Moreover, four exemplary mini-cases for the distinct company classification possibilities, namely inactive, reactive, active and proactive, are presented to illustrate the findings. Finally, a brief chapter conclusion of the overall findings is provided, wherein the two practical research questions are addressed.

Chapter four ultimately summarizes the main findings and implications of the theoretical and practical research conducted. It consists of four sections. Hereby, section one is comprised of the theoretical and practical research contribution. Subsequently, section two presents the implications of this thesis for theory and for managerial practice. This is followed by section three which entails a conclusion of the undertaken research approach and the most essential results obtained. Lastly, section four closes this thesis by outlining research limitations and presenting suggestions for further research.

2 The Structured Literature Review

This chapter contains the theoretical part of the thesis, consisting of the findings of a structured literature review. It is comprised of four sections. The first section thereby outlines the methodology of the structured literature review. Section two, in turn discusses the topic of poverty alleviation and the role of the business sector in this regard. Next, section three presents five major business strategies on poverty reduction, namely Business-as-Usual, Corporate Philanthropy, Microfinance, Bottom of the Pyramid and Social entrepreneurship, and investigates their potential impact on tackling poverty. Lastly, section four links the theme of Cross-Sector Partnership to poverty reduction as well as to the previously outlined business strategies. This chapter closes by providing a classification of companies' Poverty Alleviation Portfolios.

2.1 Methodology

This section aims at outlining the way in which the structured literature review for the theoretical part of this thesis was conducted. Matters discussed include the selection of keywords; the approach for the literature search as well as the pre-selection and the selection of the relevant articles for the literature review.

2.1.1 Keyword Selection

Concerning the keyword selection for the structured literature search, a total of 87 keyword combinations were selected by examining the themes discussed in existing research in the field of poverty and business. Table 5-1 until 5-3 of Appendix A provide a detailed overview of the employed keywords. More concretely, keywords from three different themes were combined, namely poverty, business strategies and Cross-Sector Partnerships (CSPs). The keyword 'poverty' was always maintained, whereas for the two other themes numerous synonyms and related or underlying terms were employed. As displayed in red in Figure 2-1, the focus of the literature search was set on the same three overlaps as presented in the introduction, being business strategies *and* poverty; partnership *and* poverty as well as business strategies *and* poverty *and* partnership. Due to the particular focus on the topic of poverty, the overlap between business strategies and partnerships is consequently neglected.

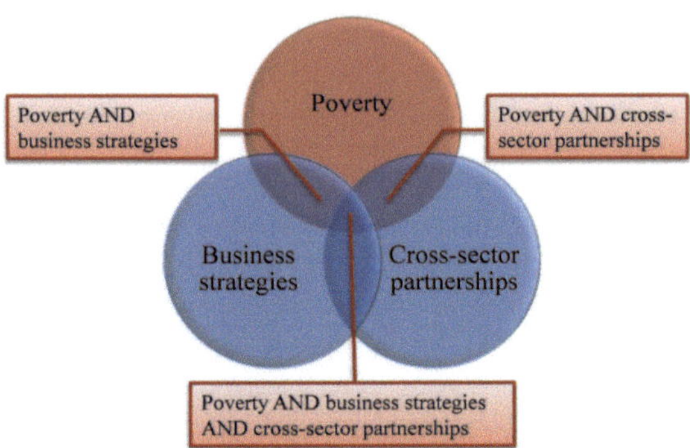

Figure 2-1: Conceptualization of Keyword Combinations Employed

2.1.2 Literature Search

The literature search of the 87 selected keywords was performed in four databases: EBSCO, Abi Inform/Proquest, Science Direct and Scopus. Thereby, a twofold search was conducted by searching in both the abstract and in the full text of the articles. In doing so, the following search criteria per database were employed:

In EBSCO a data range from year 2000 until 2011 was selected. As search functions solely 'scholarly journals' were chosen and the search was performed both in the 'author-supplied abstract' and in 'all text'.

In Abi Inform/Proquest the same data range from year 2000 until 2011 was employed. Moreover, the 'advanced search option' was checked and the search options 'peer reviewed', 'article' and 'scholarly journal' was ticked. The search was performed in the 'abstract' and under the 'all field + text' option.

In Science Direct once again a data range from year 2000 until present was used and the search was specifically conducted as a 'journal search'. In addition, 'all journals' were selected as the source and the document type was limited to solely 'articles'. Besides, the following disciplines were chosen: 'business, management and accounting', 'economics, econometrics and finance' and 'social sciences'. Similarly, the search was performed both in the 'abstract, title, keyword' category and in the 'full text'.

In Scopus the same data range from year 2000 until present was employed. Furthermore, the search was limited to 'article' as document type and to 'social science' as discipline. Ultimately, the search was again conducted in the 'abstract title, abstract, keyword' as well as the 'all fields' category.

In Table 2–1 and 2–2 the preliminary results for the searches per database and per abstract or all text are presented. Hereby, it can be seen that EBSCO was the database which yielded the most results, being followed by Abiform/Proquest and Scopus. Science Direct, ultimately, displayed the smallest number of articles. Moreover, the total articles found equaled 237,246 whereby the abstract search revealed 7,500 results whereas the all text search led to a total of 229,746 articles.

Results per database	All text	Abstract	Total
Ebsco	93,293	1,337	**94,630**
Abi Inform/Proquest	54,322	1,454	**55,776**
Science Direct	37,815	523	**38,338**
Scopus	44,316	4,186	**48,502**

Table 2-1: Structured Literature Search Results per Database

Search Results	Total
Abstract search	7,500
All text search	229,746
Both searches	**237,246**

Table 2-2: Overall Literature Search Results

2.1.3 Pre-Selection of Relevant Articles

Concerning the pre-selection of the articles, first, a brief screen of the 237,246 results was undertaken. This was conducted in order to identify articles which would be exported via Refworks to an overview Excel sheet and then would be further analyzed on their relevance by reading the abstract. Accordingly, articles were pre-selected in the following manner:

First of all, several keywords which did not yield any results were excluded, being 'strategic philanthropy', 'pro-poor marketing', 'equitable supply chains', 'fair supply chains', 'responsible Foreign Direct Investment' and 'ethical Foreign Direct Investment'. Next, several keywords which yielded a total (with respect to the abstract search within all four databases) of more than 500 articles were excluded as well, due to the fact than an overly large number of search results indicates the existence of a vague keyword. These keywords were 'standard', 'employment' and 'linkages'. However, it was ensured that relevant topics were nonetheless covered due to additional searches with other keyword combinations or within Google Scholar. For the remaining keyword combinations, all obtained articles from all four databases within the abstract search were preselected and exported to Excel. These selected articles and their keyword combinations are displayed in yellow in Table 5-1 and 5-2 of Appendix A. Additionally, if the sum of one keyword combination in all four databases was below ten, the results for the all text search were taken into account. In turn, if the results of this all text search yielded 10 or fewer article (per database), these articles were exported to Excel as well. An overview of the pre-selected articles based on their keyword combinations is exhibited in orange in Table 5-1, 5-2 and 5-3 of Appendix A.

Overall, a total of 2,703 articles were selected for further analysis. Hereby, as displayed in Tables 2-3 and 2-4, Scopus once again yielded the greatest number of results, followed by Abi Inform/Proquest and EBSCO. Science Direct had the least number of pre-selected articles. The total number of all text search articles equaled 235 whereas the sum of all selected abstract articles was 2,468.

Results per database	All text	Abstract	Total
Ebsco	49	456	**505**
Abi Inform/Proquest	60	290	**350**
Science Direct	69	201	**270**
Scopus	57	1,521	**1,578**

Table 2-3: Pre-Selected Literature Search Results per Database

Search Results	Total
Selected all text search articles	235
Selected abstract search articles	2,468
Overall selected articles	**2,703**

Table 2-4: Overall Pre-Selected Literature Search Results

2.1.4 Selection of Relevant Articles

With respect to the final selection of relevant articles, the first step encompassed the removal of duplicates, which resulted in 1,796 relevant articles. Afterwards, the abstracts of these 1,796 articles were read in order to identify relevant articles, which were within the scope of the thesis. These relevant articles would thus be read in depth. Articles were ultimately selected based on one or more of the following characteristics:

- The paper's main focus is on poverty or poverty alleviation
- The paper's main focus is on the role of business or MNCs with regard to poverty alleviation
- The paper's main focus is on approaches/strategies/initiatives of businesses or MNCs which address poverty
- The paper's main focus is on the role of Cross-Sector Partnerships which involve the private sector and address poverty
- The paper's main focus is on a combination of the three areas: poverty (alleviation), business strategies and Cross-Sector Partnerships

This final selection of articles consequently resulted in a total of 163 reviewed publications. Hereby, the literature review was based on 115 scientific articles. These included the results of the structured keyword search analysis in addition to further literature searches, for instance via Google Scholar or relevant articles mentioned within other articles. Besides, 48 books, reports and other publications identified through the structured literature review and additional searches also formed the basis of the literature review.

An overview of the characteristics of the 115 reviewed scientific articles can be found in Figure 5-1 of Appendix A as well as in Table 2-5. Hereby, Figure 5-1 reveals that the chosen articles were overwhelmingly published in recent years with solely 6 publications prior to the year 2000 and the majority of articles originating within the last 5 years. Besides, Table 2-5 aims at providing a brief overview of some of the originating journals, being selected based on frequency. The results indicate that the selected articles steam from reputable journals within a diversity of academic fields, including business-society management and business ethics; development studies; public administration and policy; or international management and business studies. Moreover, articles within the field of entrepreneurship, innovation, psychology, economics or distinct business disciplines, such as strategy, marketing, finance or accounting, were reviewed. Hence, it is revealed that research on business and poverty falls into numerous academic fields, with development and business-society management being most relevant.

Journal	Number of Reviewed Articles
Business-Society Management and Business Ethics	
Journal of Business Ethics	7
Business and Society	3
Corporate Governance	3
Greener Management International	2
Business Ethics Quarterly	1
Development Studies	
Development and Practice	5
Gender and Development	2
Journal of International Development	2
Enterprise Development and Microfinance	2
World Development	2
Community Development Journal	1
Public Administration and Policy	
International Affairs	3
Public Policy Research	1
International Management and Business Studies	
Journal of World Business	4
Academy of Management Review	3
Business Horizons	2
Journal of Management	2
Harvard Business Review	1
Entrepreneurship and Innovation	
Entrepreneurship Theory and Practice	2
Journal of Social Entrepreneurship	1
Stanford Social Innovation Review	1
Psychology	
Journal of Personality and Psychology	1
Economics	
The Review of Economics and Statistics	1
Business Disciplines	
Journal of Consumer Marketing	2
Finance and Development	1
Strategy and Business	1
Accounting, Auditing and Accountability Journal	1
Additional Journals and Other Academic Fields	
Articles from other journals	98

Table 2-5: Journal Sources of the Selected Articles

2.2 Poverty Alleviation and the Role of Business

This section aims at providing an overview of the overarching role of the private sector with regard to poverty alleviation. First, conceptualizations of poverty and its alleviation are briefly presented. Afterwards, the emerging role of business, in general and Multi-National Corporations (MNCs), in particular, is highlighted and the potential contributions and limitations of companies for tackling poverty are discussed. Ultimately, a conclusion of the findings is drawn.

2.2.1 Conceptions of Poverty

In order to further discuss approaches to poverty alleviation and the role of business, a sufficient understanding of poverty itself is required. This subsection therefore contains a brief overview of the current state of poverty in the world followed by a provision of the most essential conceptualization of poverty.

The Current State of Global Poverty

Poverty can be regarded as one of the most pressing and essential social issues at present. The World Bank estimates 47 percent of the world's population, approximately 2.5 billion people, live on less than $2 per day out of which roughly 27 percent, about 1.4 billion, survive on less than $1.25 per day (World Bank, 2011). In related terms, Sachs (2008, p. 6) states that 1 billion people "...remain trapped in extreme poverty unrelieved by economic growth." Moreover, the overall number of chronically poor has increased to 420 million people in the past decade and it has been argued by Prahalad and Hammond (2002) that the poor will face substantial difficulties in overcoming their poverty traps if current conditions prevail.

Whereas overall poverty rates have decreased slightly over the last years, progress appears to be unequally distributed. Particularly, in sub-Saharan Africa poverty remains substantial and rates are generally rising (United Nations, 2011). Besides, as of 2010, the wealthiest one per cent of people in the world earns an income equal to the total combined income of the poorest 57 per cent (Gore, 2010). Furthermore, the number of working poor remains considerable with roughly 20 per cent of all workers living in extreme poverty (United Nations, 2011). To summarize, even though estimates diverge slightly, dependent on the concrete approach employed to measure poverty, the results remain striking: Poverty remains widespread and its alleviation thus is of crucial importance.

Poverty Conceptions

However, before addressing approaches towards poverty alleviation, it is essential to clarify conceptions of poverty alleviation. Academic research on the nature of poverty has been characterized by ambiguity and disagreement, accompanied by a multitude of poverty definitions (Osmani, 2003). Figure 2-2 consequently displays key distinct dimensions of poverty which are explained subsequently.

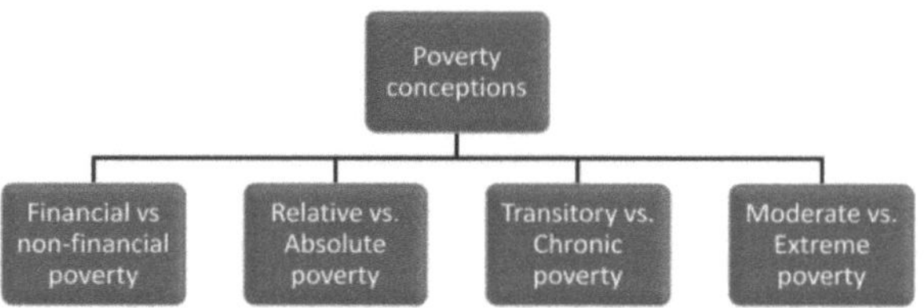

Figure 2-2: Dimensions of Common Poverty Conceptions

One primary distinction of conceptions can be made according to whether poverty is defined in economic or in non-economic terms. Hereby, one traditionally dominant conception conceptualizes poverty in *economic terms*, being a "...pronounced deprivation in well-being" (World Bank, 2011, p. 1). Well-being in this context is primarily based on the income level. This implies that the poor possess an income and consumption level which is insufficient to reach a predetermined minimum threshold. The above presented poverty conception was frequently criticized as being too narrow and was thus replaced by the established capability poverty definition. Poverty is thus perceived as a lack of basic capabilities of a person in order to function effectively in society and to live an enjoyable life. These capabilities encompass *non-economic dimensions*, such as inadequate nutrition, insufficient health and sanitary conditions or a lack of education (Rauflet, Berranger and Aguilar-Platas, 2008). Although the capability poverty definition might be more inclusive, it has nonetheless been second-guessed due to its complexity and abstraction in measurement. This has resulted in the fact that thus far no unitary accepted poverty definition can be found, with conceptions diverging according to the focus of studies (Osmani, 2003).

In addition, several other poverty-related terms need to be clarified. First, a distinction has to be made between absolute and relative poverty. *Absolute poverty* refers to "...subsistence below minimum, socially acceptable living conditions, usually established based on nutritional requirements and other essential goods" (Osmani, 2003, p. 2). In contrast, *relative poverty* "...compares the lowest segments of a population with upper segments, usually measured in income quintiles or deciles" (Osmani, 2003, p. 2) and therefore considers poverty primarily in terms of inequalities. Furthermore, chronic and transitory poverty conceptions need to be clarified. Hereby, *chronic poverty* refers to long-term poverty whereas *transitory poverty* encompasses temporary poverty conditions usually a consequence of external shocks (Osmani, 2003). Ultimately, if poverty is measured in economic terms moderate poverty and extreme poverty can also be distinguished. *Moderate poverty* relates to an income between \$1.25 and \$2 dollars a day while *extreme poverty* implies that a person lives of less than \$1.25 a day (Kelley, Werhane and Hartman, 2008).

As displayed above, conceptions of poverty are numerous and as a result, a general consensus has emerged among academia to view poverty as *multidimensional* in nature (Devajan, 2002). Examples of common poverty dimensions include income, health, education, nutrition and community participation (Hotel, 2005). These dimensions are interrelated and are best examined jointly in order to ensure a suitable conceptualization of the nature of poverty.

To summarize, poverty can be conceptualized in numerous ways with the employed definition of poverty consequently determining the dimensions being prioritized in its measurement. However, research has demonstrated that these different units of analysis on poverty tend to converge in terms of the overall picture presented (von Maltzahn and Durrheim, 2008). In this thesis, the multidimensional nature of poverty is considered even though, due to the business perspective taken, the economic dimension of poverty is still the most pivotal one in proceeding discussions. Hence, the following *poverty* definition by the Council of the European Union (2004, p. 8) will be employed:

"People are said to be living in poverty if their income and resources are so inadequate as to preclude them having a standard of living considered acceptable in the societies in which they live. Because of their poverty, they may experience multiple disadvantages through unemployment, low income, poor housing, inadequate health care and barriers to life-long learning [...]. They are often excluded and marginalized from participating in [economic, social and cultural] activities that are the norm of other people."

2.2.2 Approaches to Poverty Alleviation

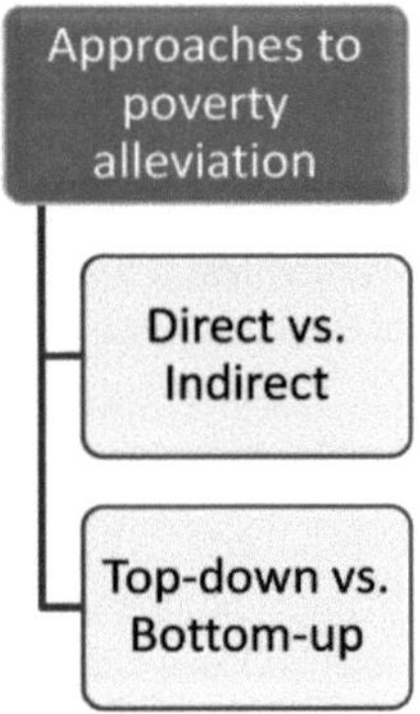

Figure 2-3: Distinct Approaches to Poverty Alleviation

Due to the widespread presence of poverty, its successful alleviation has been regarded as of vital importance but at the same time represents "...one of the greatest challenges of our time" (Viswanathan, Sridharan and Richtie, 2008, p. 228). *Poverty alleviation* can be defined as "...the improvement of well-being of poor people [...]. Well-being [in this context] encompasses income-consumption, human development, health, adequate nutrition, shelter and participation of the individual in society. Well-being thus explicitly includes [...] active participation and benefits in economic and social activities." (Hotel, 2005, p. 4). A milestone in the field of poverty alleviation has been the establishment of, and the international commitment to, the Millennium Development Goals (MDGs) in 2000, which contain the target of halving poverty by 2015 (Fukuda-Parr, 2004). As a result of the multitude of poverty conceptions, poverty alleviation approaches are also plentiful with diverging focal points. This subsection consequently presents an overview of the most common approaches to poverty alleviation in order to ultimately facilitate an understanding of the role business can play in poverty alleviation. An overview of the poverty alleviation approaches discussed can be found in Figure 2-3.

Direct and Indirect Poverty Alleviation Approaches

With regard to poverty alleviation approaches, first of all, a distinction has to be made between direct and indirect poverty reduction techniques. To begin, the *direct approach* on poverty alleviation focuses on the social, non-financial poverty dimensions and is concerned with social welfare issues, including education, health and housing, being directed particularly at marginalized groups in society (Eder and Öz, 2008). In contrast, the *indirect approach* is foremost economic in nature and focuses on the potential of the market and economic growth for reducing poverty, for instance through enhanced employment opportunities and increased income levels (Eder and Öz, 2008).

With regard to their effectiveness, it has been stated that both strategies can be successful in addressing poverty. Furthermore, both approaches are essential, in order to successfully tackle poverty, because of their interrelatedness. Poverty, due to its multidimensional nature, requires a multidimensional approach to poverty alleviation. The different approaches thus complement and not substitute one another (Eder and Öz, 2008).

Top-Down and Bottom-Up Poverty Alleviation Approaches

Next, top-down and bottom-up poverty alleviation approaches can be distinguished. In *top-down approaches*, the poor are regarded as passive recipients of development assistance, with their needs being presumed by the implementing organizations in the developed world without consultation of the poor (Osmani, 2003; London, 2007). Hence, the poor are considered as "victims that need to be helped" (McKague et al., 2004, p. 49) instead of individuals with skills and assets that should be nurtured. In contrast, *bottom-up approaches* are built on the principle of participation, in order to actively involve the poor and consider their local needs and opinions in the design of poverty alleviation strategies (Seelos and Mair, 2004).

Regarding the effectiveness of both approaches in terms of poverty alleviation, criticism of the top-down initiatives have accumulated. While top-down approaches can be successful "...in some instances, [...] this has not been the norm" (McMullen, 2011, p. 206) due to frequent negligence of the real needs of the poor. The main consensus has thus been that poverty alleviation strategies which actively include the poor are regarded as more sustainable due to greater levels of self-sufficiency and lower dependence on external aid over the long run (McMullen, 2011). Empowerment, inclusion and participation of the poor are considered crucial during all phases of poverty reduction strategies (Osmani, 2003). Overall, a general shift has recently taken place in the field of development away from top-down approaches towards demand-driven, bottom-up strategies for tackling poverty (Osmani, 2003; van Empel, 2008).

To conclude, poverty alleviation strategies are plentiful, being based on distinct approaches due to diverging underlying poverty conceptions. In general, direct and indirect poverty approaches are both regarded as necessary in order to successfully tackle poverty, due to the fact that the undertakings complement each other. Besides, bottom-up strategies are considered more effective in terms of poverty alleviation than top-down initiatives. As demonstrated in later sections, business approaches on poverty alleviation can employ top-down or bottom-up activities. They can also be direct or indirect in nature, although, they are oftentimes rather indirect than direct (Raufflet, Berranger and Aguilar-Platas, 2008). In terms of effectiveness for addressing poverty, market-based poverty alleviation approaches would consequently be preferred, if they are bottom-up in nature while addressing numerous poverty dimensions at once and thereby containing both direct and indirect aspects.

2.2.3 The Failure of the Societal Spheres and the Shift in Development Orientation

This subsection discusses the failure of the societal spheres and the consecutive shift in development orientation. The field of development and poverty alleviation has traditionally been regarded as being the primary responsibility of the government in addition to the nongovernmental sector (Farias and Farias, 2010). Hence, poverty alleviation was defined as an *interface issue* between the state and civil society, as displayed in Figure 2-4 (van Tulder and van der Zwart, 2006; Karnani, 2007a). Hereby, approaches to poverty alleviation have been primarily top-down in nature and they encompassed both direct and indirect initiatives. The decisive factor has been the fact that these undertakings were generally initiated by the state and NGOs as primary actors.

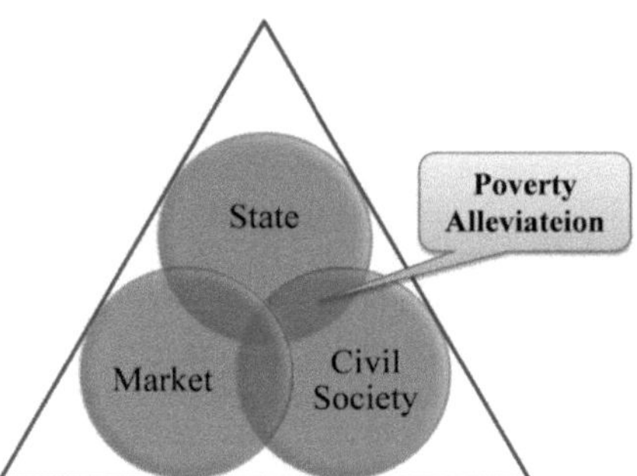

Figure 2-4: The Traditional Position of Poverty Alleviation in the Societal Triangle (based on van Tulder and van der Zwart, 2006)

However, after more than 60 years of development assistance and 2.3$ billion spend on aid to developing countries, despite certain success stories, the problem of poverty remains widespread (Karher, Iyani and Shannon, 2007; London, 2007). In this regard, certain failures of the societal spheres on the way in which development initiatives on poverty alleviation are currently executed, can be highlighted.

The Failures of the Societal Spheres

Concerning the state, *governmental failure* can be attributed due to several reasons which hinder the effective contribution of governmental efforts to development and poverty reduction. First of all, state actors are highly characterized by bureaucracy and low efficiency. Moreover, corruption is widespread (Karher, Iyani and Shannon, 2007). Additionally, with regard to poverty alleviation, a substantial amount of initiatives have been undertaken in a top-down fashion with restricted emphasis placed on local involvement, resulting in low effectiveness and high aid dependency (Kolk, van Tulder and Kostwinder, 2008). Moreover, resources were by trend spent inadequately and, as a cause of misallocation, unable to sufficiently improve conditions for the poorest of the poor (Devarajan, 2002). As noted by McCord (2004, p. 3): "To date, public work programs have not significantly stimulated local economic development, a prerequisite for the creation of sustainable job opportunities [...] Public work programmes do not necessarily [...] move participants out of poverty, but offer a temporary respite, reducing the depth of poverty during the period of employment, and they do not [...] offer sustainable livelihood improvement." Governmental approaches have thus been frequently ineffective and short-term in nature, unable to ultimately lead to self-sufficiency and sustainable development.

With respect to *civil society failure,* several aspects need to be outlined. First, despite their good intentions, NGOs are generally restricted by severe resource constraints, in particularly a lack of financing due to insufficiently gathered funds and donations (Dollery and Wallis, 2001). Beside, of these already limited financial resources, the amount, in actuality, is only partially used on the initial purpose due to substantial costs (Karher, Iyani and Shannon, 2007). Thus, the impact that NGOs can have on poverty alleviation overall is relatively constrained. Moreover, NGOs have been characterized by deficits in efficiency and effectiveness within their operations (Kolk, van Tulder and Kostwinder, 2008). Lastly, a major point of criticism has been the fact that their undertakings have frequently resulted in aid dependence instead from of a self-sufficient approach towards combating poverty (Karher, Iyani and Shannon, 2007).

The Shift in Development Orientation

As a result of the previously outlined deficiencies in the field of development and poverty reduction, new approaches have been sought. One emerging trend was the formation of Cross-Sector Partnership to address issues, such as poverty, which are beyond the scope of a single actor. This development is further outlined in section 2.4. The second evolving movement was the increased involvement of the business sector in development, being accompanied by the enhanced establishment of market-based approaches to poverty reduction (McKague et al., 2004; Frynas, 2008). Hence, as displayed in Figure 2-5, poverty alleviation shifted from being an interface issue towards being a *growth regime issue,* which requires the involvement of all three societal actors – including the market – for its successful combat (Sharp, 2006; ‚Karnani, 2007a; van Tulder and van der Zwart, 2006).

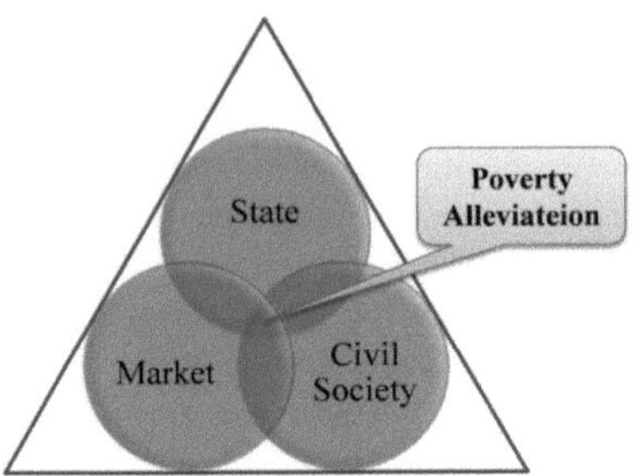

Figure 2-5: The Evolved Position of Poverty Alleviation in the Societal Triangle (based on van Tulder and van der Zwart, 2006)

2.2.4 The Relevance of the Business Sector for Poverty Alleviation

In particular within the last decade, the importance of the business sector, in general, and Multi-National Corporations (MNCs), in particular, on poverty alleviation has frequently been highlighted in academic research. Subsequently, this subsection will further dive into this topic and outline the assumed role of the business sector with regard to tackling poverty.

The Importance of the Business Sector for Poverty Alleviation

In recent debates on development and poverty alleviation, the emerging relevance and evolved role of the private sector has regularly been emphasized (Frynas, 2008; Stefanovic, 2007; Sharp, 2006). For instance, Barney (2003, p. 263) states that "...the private sector is increasingly seen to have a role to play in helping to meet international development targets on poverty reduction." In addition, Schwittay (2011, p. 73) announces the 'marketization of poverty' by reasoning that "...poverty is reconceptualized as a problem that can be solved by market mechanisms." To provide an overview of the variety of claims and statements made in this regard, Table 5-4 of Appendix B displays a selection of relevant quotes in this context. From these citations, it can be noted that generally the majority of authors seems to agree on the fact *that* the private sector has an important role to play in combating poverty. However, as outlined repeatedly, on the questions of *what* this role and the potential contributions look like and *how* business initiatives on addressing poverty should be designed, opinions remain contradictory.

The Importance of MNCs for Poverty Alleviation

Focus in discussions of the role of business has been placed particularly on MNCs as unit of analysis. Accordingly, Table 5-5 in Appendix B displays literature quotes of the importance of MNCs for poverty alleviation. The underlying reason is the powerful role which MNCs nowadays play in society and in development. To illustrate this, the levels of yearly Foreign Direct Investment (FDI) currently exceed the amount of yearly Official Development Assistance fivefold (McFalls, 2007). Besides, approximately one fourth of all global activity can be attributed to MNCs (McFalls, 2007). Thus, the scope of MNCs is substantial, their presence is widespread and hence their impact on poverty could be enormous, if approached in the right manner. Whereas discussions have contained both, the potential positive and negative consequences on MNCs' operations, this thesis aims at focusing solely on the potential *positive* contribution which MNCs can make by regarding their 'poverty-alleviation potential' (Kolk and van Tulder, 2006, p. 790). However, it has to be emphasized that the impact of a firm's operation on poverty is best considered in terms of the net effect, which refers to the total positive effect minus the total negative effect. Hence, entirely leaving out the negative implications and points of criticism concerning the business approaches on poverty would result in an incomplete picture. Consequently, the negative aspects of the selected business initiatives will be briefly discussed as well when examining these approaches in latter sections.

To summarize, the main findings from the quotes of Table 5-4 and 5-5 in Appendix B state that the private sector's engagement is vital for poverty alleviation and that in particular large MNCs have an important role to play by contributing to successfully combating poverty.

2.2.5 Contributions and Limitations of Involving the Business Sector

After previously outlining the overall importance of business for poverty alleviation, one wonders what the potential contributions and limitations of involving the private sector into development may be. This matter is discussed in this subsection.

Potential Contributions of the Business Sector to Poverty Alleviation

Concerning the extent to which business can possibly contribute to poverty alleviation, an array of opinions among researchers can be found. Table 5-6 and 5-7 in Appendix B consequently presents relevant quotes on the potential business contributions and limitations to combating poverty. Considering the contributions, two possible underlying dimensions can be distinguished, namely the financial and the non-financial impact of the involvement of businesses into poverty reduction, as displayed in Figure 2-6.

Financial Contributions of the Business Sector to Poverty Alleviation

To begin with, the business sector can positively contribute to poverty alleviation through *financial gains* for the poor. Such an indirect contribution is presumed to occur mainly via economic growth, being a by-product of business engagement in a developing country. Such economic growth can ultimately be beneficial for the poor, if it is inclusive and equitable so that its benefits reach the poor as well (van Tulder and da Rosa, 2010). Other possible financial benefits obtained include the general provision of (enhanced) job opportunities which might be accompanied by better working conditions and employment opportunities with higher income (UNDP, 2004). Furthermore, financial access to the poor has been highlighted as a favorable consequence of business operations (Yunus, Moingeon and Lehmann-Ortega, 2010). Lastly, taxation payments by companies to governmental actors might transform into benefits for the poor, if spent subsequently on poverty-related initiatives (Newell and Frynas, 2007; Singer, 2006).

Non-Financial Contributions of the Business Sector to Poverty Alleviation

In contrast, *nonfinancial business contributions* could occur through capacity building activities and employee training which might result in enhanced employee skills and better job prospects (Singer, 2006; Kolk and van Tulder, 2006). Besides, benefits could be obtained in form of empowerment, community involvement and social inclusion which can all enhance the self-esteem of poor citizens (UNDP, 2004). Finally, greater freedom and choice, more favorable living situations as well as improvements in the external environment through business initiatives could eventually increase the quality of life of the poor and lead to more favorable environmental conditions (Onaga and Ogbalu, 2008).

Figure 2-6: The Potential Contributions of the Business Sector on Poverty Alleviation

Limitations of the Business Sector and the Evolved Societal Roles

Overall, while the debate on the exact extent of the potential contribution of business to poverty reduction remains inconclusive, a general consensus has emerged, that business actions alone will be insufficient to combat poverty (Edar and Öz, 2008). Business initiatives should therefore be regarded as a *complement* to, and not as a substitute of, existing development initiatives which address poverty (Merino and Valor, 2011). Undertakings of other societal actors are still highly required. Government actors are needed in order to address the social dimensions of poverty and thereby should provide the basic foundations, in terms of education, health or infrastructure, which are aspects largely unaddressed by business initiatives. Moreover, the government needs to create an environment that enables and monitors business activities (UNDP, 2004). Besides, *market failures* could also occur as well due to the fact that certain societal groups – often the poorest of the poor – may be excluded from particular business undertakings, even if targeted at the field of poverty alleviation (Sharp, 2006). Therefore it is vital that NGOs and governments engage in targeted initiatives to support these marginalized groups of society in sustainable manners.

Another limitation and frequently mentioned point of criticism represents the foremost focus on the business impact of private sector engagement in the field of poverty alleviation. While certain claims are made on the potential contributions of business initiatives on poverty reduction, widespread empirical evidence is frequently lacking (Lyon and Bertotti, 2007). Moreover, this possible contribution tends to be contingent on the precise business approach undertaken and therefore more distinct analysis of the impact on poverty alleviation per business approach is required (Newell and Frynas, 2007). This matter will therefore be analyzed in the subsequent sections

All in all, the literature analysis has revealed that the potential contributions of private sector engagement in poverty reduction *could* be substantial. However, whether this potential is actually realized depends on the manner of undertaking and the type of strategy pursued.

A Concluding Remark

To summarize, this section has provided an overview of the theme of poverty alleviation and the role of business therein. In this regard, several concluding statements can be made. To begin, poverty is a highly complex and controversially defined issue due to the fact that it is multidimensional in nature. Moreover, approaches to poverty alleviation are diverse as well and take up different angles. Important distinguishing factors are whether these approaches are characterized as direct or indirect and whether they classify as top-down or bottom-up strategies. Next, concerning the field of development and poverty reduction, past approaches initiated by governments and NGOs have frequently turned out to be unsuccessful. This can be attributed to several market and civil society failures. As a result, a shift in development orientation has been taken place which highlights the importance of regarding the business sector, in general, and MNCs, in particular, as essential actors for tackling poverty. However, academic opinions on the extent to which companies can contribute to poverty alleviations are ambiguous. Hereby, a distinction needs to be made between numerous financial and non-financial gains obtained by the poor through business initiatives. Therefore the precise role and potential contribution of companies towards combating poverty is contingent on the concrete business approach employed. The next section consequently presents several employed business strategies and discusses their eventual impact on poverty alleviation.

2.3 Business Strategies on Poverty Alleviation

This section analyzes five predominant business strategies on poverty alleviation, namely Business-as-Usual, Corporate Philanthropy (CP), Microfinance, Bottom of the Pyramid (BOP) and Social Entrepreneurship (SE). It is structured as follows. First, a classification of important, related terms is presented to facilitate further understanding of the subsequently discussed business approaches to poverty alleviation. Afterwards, in turn, each of the previously stated business strategies is presented and evaluated. Ultimately, a classification of these five strategies is provided based on its effectiveness in addressing poverty.

2.3.1 Definition of Terms

This subsection presents an overview of essential terms for an understanding of the area of business and poverty. In particular, an in-depth analysis of the notion of Corporate Social Responsibility (CSR) is provided. In this regard, a CSR classification framework developed by van Tulder and van der Zwart (2006) is introduced. Ultimately, an introduction to the subsequently discussed business strategies on poverty alleviation is provided.

Essential Terms

The engagement of companies in matters related to society and sustainability has been classified using a wide variety of terms. Hence, an overview and definition of essential terms is crucial to facilitate understanding of the themes discussed. Important terms in this regard and their employed definitions include the following:

- *Corporate Social Responsibility:* "An umbrella term for a variety of theories and practices all of which recognize the following:
    a) That companies have a responsibility for their impact on society and the natural environment, sometimes beyond legal compliance and the liability of individuals;
    b) That companies have a responsibility for the behavior of others with whom they do business with; and that
    c) Business needs to manage its relationships with wider society, whether for reasons of commercial viability, or to add value to society." (Blowfield and Frynas, 2005, p. 503).
- *Sustainable Development:* "Development that meets the needs of current generations without compromising the ability of future generations to meet their needs and aspirations." (WECD, 1987, p. 43).
- *Triple-Bottom Line:* "Corporate success is evaluated not only through the conventional bottom line involving financial results but also through the bottom lines of economic and social performance." (Simola, 2007, p. 131).
- *Sustainable Global Enterprise:* "The social component of the Triple-Bottom Line is now also seen as an integral and interwoven element of a pioneering and compelling approach to international business known as Sustainable Global Enterprise. Sustainable Global Enterprise is based on the premise that MNCs face numerous, significant challenges for which they must develop innovative solutions." (Simola, 2007, p. 131).
- *Inclusive Business:* "Making low income communities part of the core business of companies, as an option for significant and sustained impact on poverty." (World Bank website, 2008 as cited in van Tulder, Fortainer and da Rosa, 2011, p. 4).
- *Inclusive and Dynamic Business:* "Some call it BOP, others *pro-poor business,* yet others call it creating *sustainable livelihoods*. Whatever the name, the primary focus remains the same – [to] serve the needs of the poor through profitable commercial activities." (Onaga and Ogbalu, 2008, p. 137).

- *Inclusive Markets*: "Create systems of exchange that offer expanded choice and opportunities for the poor and produce outcomes that benefit the poor." (UNDP, 2008, p. 15).
- *Creative Capitalism*: "Combining the pursuit of profits with services to those who lack the means to pay for basic needs." (Goldsmith, 2011, p. 15).
- *Stakeholder Capitalism:* "A view that holds responsibilities of the firm extends beyond shareholders to a number of interested parties. As these stakeholders have an interest in the firm, it must recognize, if not heed, these interests." (Stoner and Wankel, 2007, p. 33)
- *Sustainable Livelihoods/Blended Value Approach*: "Doing business with the poor in ways that simultaneously benefit disadvantaged communities and benefit the company." (WBCSD, 2004, p. 6).
- *Corporate Citizenship*: "Companies should be responsible members of the communities in which they operate." (van Tulder and van der Zwart, 2006, p. 193).

As revealed through the definitions provided above, a variety of terms are used to discuss the engagement of companies in sustainable and effective initiatives on addressing social issues, such as poverty. For simplicity reasons, these distinct company initiatives are classified in this thesis under the umbrella term of CSR. In order to better understand the variety of existing CSR approaches and to see how the distinct business initiatives could be grouped, a classification framework of distinct CSR approaches is presented and consequently applied to the examined business approaches on poverty alleviation.

The CSR Classification Framework

Van Tulder and van der Zwart (2006) have developed a framework which classifies distinct CSR approaches in terms of their sustainability towards addressing societal problems. This categorization can also be applied by focusing on one concrete issue, such as poverty alleviation. Hence, as shown in Figure 2.7, four distinct CSR notions (on poverty alleviation) can be distinguished, being inactive, reactive, active and proactive.

First of all, an *inactive CSR approach* implies that a firm is primarily preoccupied with profit maximization and does not actively aim at combating societal issues, such as poverty alleviation. Hence, the operating notion would be described as 'corporate self-responsibility' where a firm focuses on efficiency dimensions and goes only as far in addressing societal matters as required by law. Companies subsequently employ an inside-in perspective by aiming at 'doing things right' and 'doing well', with performance being measured primarily in financial terms (van Tulder and van der Zwart, 2006).

Second, a *reactive CSR approach* falls under the notion of 'corporate social responsiveness'. Hereby, companies are motivated by external stakeholder pressures for more responsible business operations. Emphasis is therefore placed on minimizing inefficiencies as well as the negative impacts of noncompliance in order to 'do not do things wrong'. This relates to an outside-in perspective under the belief of 'doing well and doing good'. Thus, poverty matters are addressed in a quick-fix manner solely to the extent as it is externally required by stakeholders to avoid reputational damage (van Tulder and van der Zwart, 2006).

Third, an *active CSR approach* is described as conventional 'corporate social responsibility' where societal action is motivated foremost by ethical values and moral implications. The importance under this perspective lays on operating with integrity by 'doing good'. The responsibility of combating poverty is regarded as serious and active engagement in related approaches is willingly sought. The underlying notion of such an inside-out perspective can best be described as 'doing the right things'. However, – despite good intentions – these approaches might not be sufficient to eliminate poverty over the long run due to a lack of self-sufficiency of the pursued undertakings (van Tulder and van der Zwart, 2006).

Fourth, a *proactive CSR approach* correlates to 'corporate societal responsibility' where the effectiveness dimension is highlighted. Hence, poverty alleviation is regarded as an essential business responsibility, but it should be pursued in a sustainable and effective manner which frequently implies the existence of a business case for self-sufficiency. In turn, an in-outside-in/ out perspective is employed which aims for 'doing the right things right' as well as 'doing well by doing good'. Consequently, under this approach companies are presumed to be most engaged in poverty alleviating activities, by aiming at effectively tackling the roots of poverty (van Tulder and van der Zwart, 2006).

In his framework, van Tulder and van der Zwart (2006) also distinguish between two dimensions, in terms of tension fields. First, the inactive and the active approach fall under the *resource-based view*, which focuses on the internal attitude and behavior with regard to sustainability and CSR. In contrast, the reactive and the proactive approaches relate to the *stakeholder theory*, which implies an external motivation, driven by stakeholder demands and legitimacy considerations.

To conclude, the inactive CSR approach on poverty alleviation is regarded as least effective whereas the proactive approach is considered as most effective in tackling poverty in a sustainable manner, with the reactive and active approach falling in between. Consequently, proactive and active approaches are preferred over reactive and inactive as an ultimate company classification on its Poverty Alleviation Portfolio.

Figure 2-7: The CSR Approach Classification (Based on van Tulder and van der Zwart, 2006)

The Added Value of the Classification Framework

After describing the previously outlined framework, one may wonder what the purpose and added value of such a framework might be. Respectively, it shall be stressed that through a classification, approaches can be placed in the broader context which in turn enables comparison of initiatives based on certain aspects. In this case, the comparison dimension is the effectiveness of the selected approaches on poverty alleviation. The framework shall therefore serve as an evaluation basis of the initiatives.

Due to the extensive literature review employed in this thesis, it became clear that comparing or classifying distinct approaches is not a common undertaking. Except for a previous classification of poverty initiatives on the presented CSR framework by van Tulder (2010) himself, which this thesis aims at enriching, no categorization has otherwise been undertaken. As a result, the two remaining literature review chapters classify selected poverty alleviation approaches, as well as their variations and combinations, according to the presented framework.

Business Approaches on Poverty Alleviation

As previously mentioned, in the proceeding subsections several distinct business approaches on poverty alleviation are discussed and ultimately classified. The choice of approaches was derived from the literature search results and ultimate keyword choices as outlined in the methodological part. Hence, it was essential that the selected approaches were covered in sufficient depth in existing literature to facilitate the proceeding analysis. Moreover, approaches had to be applicable to MNCs and relate directly to poverty due to the focus of analysis. Respectively, a total of five business strategies have been selected, namely Business-as-Usual, Corporate Philanthropy (CP), Microfinance, Bottom of the Pyramid (BOP) and Social Entrepreneurship (SE) and additionally the concept of Cross-Sector Partnerships (CSPs) on poverty alleviation has been included as further initiative.

The structure in all following subsection on the business approaches on poverty reduction is identical as follows: First, the approach and – if existing – its variations are presented. Afterwards, prior research in the field and potential research gaps are outlined. This is followed by an overview of possible organizational characteristics and activities undertaken in that context. Next, the underlying business notions and possible business impacts of the distinct approaches are discussed. Subsequently, combination possibilities with other business strategies or through partnerships with other societal actors are examined. Subsequently, the overall impact of active engagement in an approach on the poor and on poverty is investigated. Lastly, an overarching evaluation of the benefits, limitations and criticisms per approach is provided. This analysis ultimately leads to a classification per approach and its varieties with regard to its potential effectiveness for poverty alleviation. Overview Tables on the characteristics of the selected business approaches and their variations can be found in Table 16 to 21 of Appendix E.

2.3.2 Business-as-Usual

This subsection discusses the potential impact of Business-as-Usual on poverty alleviation, primarily through the effect of Foreign Direct Investments (FDI) in developing countries.

The Approach

Business-as-Usual, as already indicated by the name, is not a poverty-focused business approach. Instead, it refers in the context of poverty to the impact of general business operations on the poor, most commonly expressed via Foreign Direct Investment in a developing country (Ayanwu, 2006). Companies do not actively pursue targeted poverty alleviation initiatives but consider combating poverty the responsibility of the government and NGOs. Thus, when asked how they contribute to society and the poor, companies will simply stress the positive impact business activities in general and FDI in particular have on economic growth which in turn might ultimately contribute to poverty alleviation (Newell and Frynas, 2007).

The Research Field

The research field of Business-as-Usual, especially concerning the impact of FDI on local communities, is widely researched. However, it prevalently investigates the potential *negative* influence which global business operations could have on developing countries and accordingly presents suggestions on how to *minimize* such negative impacts (Kolk and van Tulder, 2006). In contrast, a focus on how to maximize the possible positive contribution of FDI on development countries – as examined in this thesis – is rarely discussed. In this regard, academic research tends to highlight the indirect impact that FDI can have on poverty reduction through its triggering effect on economic growth (Nunnenkamp, 2004; Jenkins, 2005). Hereby, results generally remain inconclusive and vague due to a preponderance of theoretical reasoning and a lack of empirical testing of precise results (Lyon and Bertotti, 2007; Jenkins, 2005). An indication of the most vital literature review findings on the impact of Business-as-Usual on poverty reduction is revealed in Table 5.8 of Appendix C.

Organizational Characteristics and Activities

Due to the fact that Business-as-Usual in this context encompasses basically all business ventures, without a substantial focus on societal impact and CSR, restrictions in organizational design are limited. With regard to size, no specifications can be made even though MNCs, due to their internationalization and their sheer size, are generally more likely to expand operations into foreign countries (Karher, Iyani and Shannon, 2007). As the focus of this thesis is set on MNCs, primarily large, international firms with FDI operations in developing countries will be considered in proceeding discussions. These businesses can operate in all possible business sectors, since FDI is possible in basically any industry. FDI undertakings are generally at the core of a firm's strategic operations, being of vital importance to a firm's performance (Stoner and Wankel, 2009). However, poverty will not be considered as a core activity due to the fact that no engagement in any poverty-focused activities is pursued.

Underlying Business Notion

The underlying logic of Business-as-Usual dates back to Friedman and his notion "... that the only business of business is business." (Friedman, 1970, p. 122). This implies that companies should seek to maximize profit and thus regard shareholder concerns as their top priority. Responding to the needs of other stakeholders, such as the poor, is viewed as a value destroying activity for shareholder money (Newell and Frynas, 2007). Accordingly, companies should focus on what they do best – generating financial value – and should not occupy themselves with social activities out of their range of expertise – such as tackling poverty (Warhust, 2005). Hence, the main business benefits of a Business-as-Usual approach encompass the maximization of financial wealth which correlates to the business performance of the firm (Karher, Iyani and Shannon, 2007). Through engagement in developing countries via traditional FDI, a firm's revenue streams can be increased and its growth potential can be nurtured. Hence, companies operate as financially self-sustaining entities, and scalability is generally widely possible to the extent that profitable FDI opportunities can be discovered.

Compatibility

Overall, the compatibility of Business-as-Usual with other approaches is relatively limited in comparison to the remaining approaches. Whereas partnering is generally possible, partnerships are initiated mostly with companies and less with actors from distinct societal spheres (Ayamwu, 2006; Gifford, Kestler and Anand, 2010). It is also unlikely that such partnerships would emphasize poverty-related goals. Concerning the possibility of combining Business-as-Usual with the presented business strategies, the likelihood of combination is relatively low due to the fact that all other investigated initiatives explicitly aim at addressing poverty alleviation whereas Business-as-Usual does not prioritize poverty (Warhust, 2005).

Impact on Poverty Alleviation

The Business-as-Usual strategy views poverty in economic terms. However, it does not aim at directly addressing poverty but instead engages in normal business activities and FDI initiatives which have the potential to indirectly affect poverty reduction (Ayanwu, 2006). This possible indirect impact entails financial and non-financial aspects, with the former being more substantial. Accordingly, FDI can positively impact the financial dimension of poverty by proving income-generating employment possibilities and by stimulating economic growth through the financial inflows via investments (Ayanwu, 2006). Moreover, taxation payments from FDI can subsequently be spent on poverty-related activities (Newell and Frynas, 2007). Concerning the non-financial aspects of poverty, business activities in developing countries could enhance the quality of life of the poor by introducing novel, valuable products to the market (Newell and Frynas, 2007). Besides, technology transfer and skills training at work could occur which might contribute to capacity building among poor citizens (Ayanwu, 2006). In contrast, negative effects of this strategy on the poor include the crowding out effect of local firms, the creation of dependence as well as the race to the bottom for employment standards and wages and an according degradation of the working conditions (Newell and Frynas, 2007). Generally speaking, under the top-down Business-as-Usual notion, the poor are overall viewed as passive community members within the regions where firms operate and thus, active involvement of the poor is marginal (Hossain, 2000).

Evaluation of Strategy

Criticism of the Business-as-Usual notion on FDI – in general and in terms of poverty alleviation – is abundant (Karher, Iyani and Shannon, 2007). Whereas such an approach might be beneficial in financial term for companies, the potential positive impact on the poor is highly restricted (Gore, 2010). Concerning the argument presented previously, several shortcomings can be outlined. First of all, with respect to the potential financial gains of FDI for the poor, chances are unlikely that taxation payments will be spent fully on poverty related activities (Devarajan, 2004). Moreover, due to the fact that MNCs frequently require highly skilled workers for their jobs in developing countries, the extent of poor citizens employed is usually very low (Nunnenkamp, 2004). Hence, the outlined advantages in terms of income generation or capacity building opportunities may instead reach the wealthier segments of the population. Accordingly, products offered by MNCs are foremost targeted at the upper class consumers as well, so potential gains in quality of life through greater product variety are marginal as well (Devarajan, 2004). In particular with regard to the most highlighted effect of FDI, namely as stimulator of economic growth, it has been frequently criticized that such growth may not reach the poorest of the poor (Devarajan, 2004; Norton, 2002). Hence, social exclusion of the marginal societal groups may be worsened (McMullen, 2011). FDI might therefore rather lead to an increase in the living situation of the wealthier citizens while solely marginally contributing to the lives of the poor. Benefits are unevenly distributed and social inequalities might be even increase further (Norton, 2002). Overall, Business-as-Usual executed through FDI in developing countries has been highly criticized by researchers as being insufficient to end poverty (Merino and Valor, 2011).

The Classification

Overall, based on the previous discussion, Business-as-Usual – being executed in this regard through FDI into developing countries – classifies as *inactive poverty alleviation strategy*. The ultimate classification is displayed in Figure 2-8 and an in-depth overview of the underlying reasoning for the employed classification of Business-as-Usual can be obtained in Table 5-16 of Appendix E. Due to the fact that poverty reduction is not actively, but passively, addressed, this initiative is regarded as insufficient to end poverty. Such a strategy is no longer inadequate since business does neither exert its social responsibility nor its poverty-alleviating contribution. Instead firms stick to their usual business notion, which is narrowly focused on maximizing shareholder wealth while neglecting demands from other stakeholders and society. Whereas the indirect effect of general business operations can partially influence the financial and non-financial aspects of poverty, such gains are unequally distributed and are especially insufficient to lift the poorest of the poor permanently out of poverty.

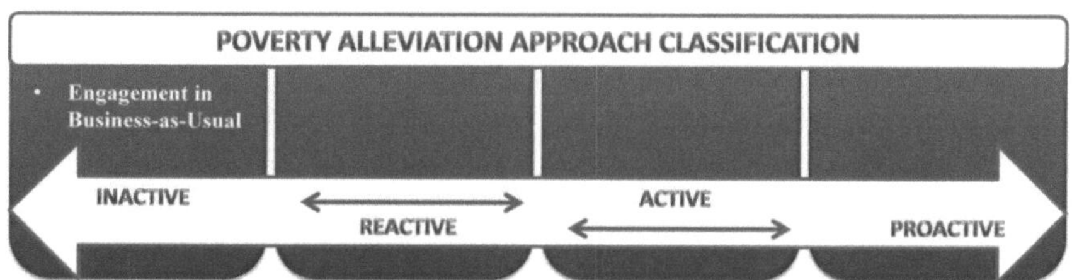

Figure 2-8: Classification of Business-as-Usual

2.3.3 Corporate Philanthropy

This subsection investigates the Corporate Philanthropy strategy in relation to poverty alleviation.

The Approach and its Variations

Corporate Philanthropy (CP) refers to beneficences provided by a firm to society in general or to a targeted marginalized group, in this case, the poor. These beneficences can be provided in many forms, including via donations, funding, sponsoring or through corporate volunteering (Montgomery, Palma and Hoagland-Grey, 2008). However, corporate donations are the most frequently examined form of CP and will be the focus of this discussion. In this regard, donations can be financial or non-financial, for instance via equipment or knowledge, and they can be provided in two manners. On the one hand a firm can donate to an external organization, usually a NGO or governmental institution, which focuses on a certain matter the company is interested in. On the other hand a firm can also establish a foundation to which internal corporate funds flow to (Leisinger, 2007). Along those lines, a distinction needs to be made among charity and philanthropy, where *charity* encompasses "...unconditional short-term relief" whereas *philanthropy* is more sustainable as it "...attempts to investigate and address the underlying causes to make a tangible positive change in the social conditions that cause the problem" (Leisinger, 2007, p. 325).

Two variations of CP need to be distinguished, namely *classical Corporate Philanthropy* and *strategic Corporate Philanthropy*. To begin, the classical notion of CP implies that the firm donates resources to a cause which does not necessarily have to be related to a firm's core operations (Leisinger, 2007). In contrast, in the case of strategic CP, a company would engage in a matter which falls into a firm's core operations and which has the potential to positively affect a company's performance (Stoner and Wankel, 2007; Tang and Li, 2009). Hereby, literature results have emphasized that poverty-related activities are more essential when they are part of a firm's core activities than when they are considered peripheral practices (Barney, 2003).

The Research Field

The field of CP and its impact on reducing poverty is highly researched within academia. Most frequently, CP efforts have been analyzed in the same manner as general aid provisions by NGOs and governments. Criticism of CP has been abundant, in terms of the short-lived nature of such assistance and the high potential aid dependency created (Leisinger, 2007). Nonetheless, the substantial financial resources of firms and their potential positive impact on development and poverty alleviation have been highlighted in discussions as well (Montgomery, Palma and Hoagland-Grey, 2008). A summary of the most relevant literature review findings on the impact of Corporate Philanthropy on poverty alleviation is provided in Table 5.9 of Appendix C.

Organizational Characteristics and Activities

With regard to the organizational characteristics for engagement in CP, generally, no restrictions in size are required. Nonetheless, due to fewer resource constraints, MNCs are more likely to undertake philanthropic activities than SMEs (Tang and Li, 2009). Besides, no limitations in terms of company sectors have to be made. As previously mentioned, strategic CP, at least partially, falls within a company's core activities while classical CP is usually not part of a firm's core activities (London, 2007).

A further distinction of the two variations is the type of activities undertaken. As already indicated above, classical CP provides financial donations, primarily to external organizations whereas strategic CP is accompanied by financial and non-financial assistance, generally being executed by an own organization (Leisinger, 2007). Employee giving and training is also more likely within strategic CP. Additionally it is presumed that the number of initiatives, in general, as well as those with a poverty focus, is more extensive in strategic CP than in classical CP.

Underlying Business Notion

The underlying business motives of engagement in classical CP are primarily external stakeholder pressure for more responsible business practices and the potential for reputational gains (UNDP, 2004; Leisinger, 2007). For the strategic CP approach, this is enriched by the potential to engage in philanthropic activities related to the company's core activities and field of expertise (Tang and Li, 2009). Hence, besides improved stakeholder relations, a strategic business notion is the driving force for active engagements. However, in both instances the financial self-sufficiency is relatively low, even though it is slightly higher for the strategic CP approach. Moreover, scalability is substantially restricted by limitations in available funding for project expansions (UNDP, 2004).

Compatibility

Concerning the potential compatibility with other approaches, likelihoods are high, in particular for partnerships with governments or local NGOs (Tang and Li, 2009; Moran, 2008). The more strategic a CP approach, the higher is the likelihood that companies could actively assist, with their field expertise, in the local implementation of a project. Furthermore, a combination with other business strategies on poverty alleviation is high as well. Particularly, compatibility with Microfinance as well as BOP and SE can be expected, especially through the provision of extra funding for social undertakings (Goldsmith, 2010). Overall, the likelihood and extensiveness of possible combinations of the approaches is presumed to be higher for strategic CP than for classical CP.

Impact on Poverty Alleviation

The impact of CP on poverty is widely discussed – in positive and in negative terms (Moran, 2008; Sasse and Trahan, 2006). Philanthropic activities can be undertaken in an array of ways and thus can address various direct or indirect dimensions of poverty, contingent of the concrete focus at hand. In similar terms, the obtained benefits of the initiatives can vary as well, ranging for instance from financial gains, such as employment, to non-financial benefits via improved living conditions. However, projects are usually executed in a top-down manner with low involvement of the poor. Instead, the poor are regarded as passive recipients of external assistance and low emphasis is placed on permanently nurturing or training the skills of the poor (Leisinger, 2007).

Evaluation of Strategy

Based on the potential impact of classical CP on poverty described above, frequent critiques emphasize the short-term, irregular and nonstrategic notion of such a poverty alleviation approach as well as its lack of focus on obtained results (Thurman, 2006; Sasse and Trahan, 2006; UNDP, 2004). Along these lines, the low involvement of the poor and the dependency, which is oftentimes created, are generally highlighted (Leisinger, 2007). Furthermore, the mssing inclusion into core business activities is criticized in the classical approach. Moreover, the potential impact of such CP – in particular when referring to the classical concept – is severely restricted by available financial funds, which will simply be insufficient to end overall poverty (Tang and Li, 2009). Since the sustainability and self-sufficiency are slightly higher for the strategic variation, criticisms are less abundant. Nonetheless, the common understanding remains that a not entirely self-sustaining business approach without involvement of the poor will still fail to lift the poor permanently out of poverty (WBCSD, 2004; Tang and Li, 2009).

The Classification

On the basis of the previous evaluation the classical CP strategy classifies as a *reactive poverty alleviation approach* whereas the strategic CP notion can be defined as an *active poverty alleviation approach*. The final classification is visualized in Figure 2-9 and an extensive classification table of CP with additional information can be retrieved in Table 5-17 of Appendix E. To summarize, it has been acknowledged that corporate philanthropic activities have the potential to address basically all dimensions of poverty and thus could significantly improve living conditions for targeted poor individuals. Nonetheless, especially in the classical variation, funds are simply insufficient to lift all poor out of misery. Moreover, the high aid dependency that is created and the low inclusion of the poor make this approach relatively unsustainable in addressing poverty on a long-term basis.

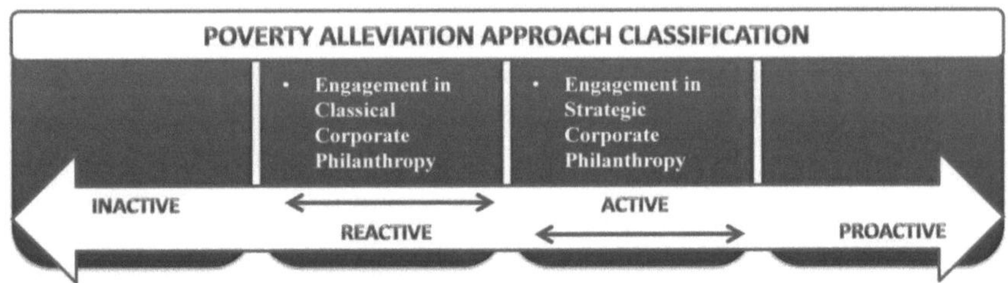

Figure 2-9: Classification of Corporate Philanthropy

2.3.4 Microfinance

This subsection discusses the business strategy of Microfinance and its potential contribution to poverty alleviation.

The Approach and its Variations

Microfinance is a concept which has been popularized by Nobel Prize winner Muhammad Yunus. The underlying logic is the fact that among the poor access to finance is highly restricted, due to low levels of credibility (Yunus and Jolin, 1999). As a consequence, Yunus aimed at providing financial services to the 'unbankable' in order to help them escape poverty (Jones Christensen, 2008).

In this regard, two terms need to be distinguished. The first one is *Microcredit* which encompasses the exact provision of small loans to the poor, as described above. In contrast, *Microfinance* is a broader term, which – besides credit provision – also offers additional services, such as insurance, housing loans or credits for durable consumer goods (Jones Christensen, 2008). Whereas Microfinance might overall increase the quality of life of the poor, Microcredit provisions for entrepreneurial activities can be regarded as a basic necessity in order to lift people out of poverty.

With regard to its variations, a narrow and a broad Microfinance approach can be distinguished. *Narrow Microfinance* relates to the exclusive provision of loans for entrepreneurial activities whereas *Broad Microfinance* also engages in supporting activities in order to foster the entrepreneurial skills of the poor to enhance their likelihood of business success (Karlan and Valdivia, 2011). Moreover, the variations can be differentiated based on their financial independence. In the provided distinction, non-self-sufficient operations which would require external funding would fall under the narrow approach, whereas self-sufficient initiatives would classify as broad Microfinance strategies (Mosley and Hulme, 1998).

The Research Field

Overall, the academic research field on Microfinance is once again characterized by ambiguity and inconsistency of opinions and empirical findings (Johnson, 2009). One explanation therefore could be the variety of Microfinance approaches, as previously outlined. Moreover, it appears as if the research outcomes are contingent on the concrete manner in which a Microfinance strategy is employed. Hence, it is important to pay close attention to key success factors for the implementation, in order to maximize the potential impact on poverty alleviation. Nonetheless, out of all five examined business strategies on poverty reduction Microfinance appears to be the most widely discussed one in academia. Accordingly, empirical findings have been most concrete with measurable – but contradicting – results in comparison to the remaining strategies explored. An overview of the most essential literature review findings on the impact of Microfinance on poverty alleviation can be found in Table 5-10 of Appendix C.

Organizational Characteristics and Activities

Several possible aspects in terms of design for Microfinance initiatives have to be stressed. To start, Microfinance can usually be initiated either as a business approach or it can be designed as a nonprofit operation (Jones Christensen, 2008). In this thesis, the focus logically remains on business-related Microfinance activities. Moreover, Microfinance can stand at the core of a company's operations as in broad Microfinance, or can be solely part of the firm's peripheral activities. In the latter case, Microfinance might even be provided by companies in terms of philanthropic, non-self-sufficient activities which would correspond to the narrow Microfinance notion (Mosley and Hulme, 1998). Concerning the possible business sectors, financial firms are more likely to engage in microfinance if it is at the core of the operations although generally all sectors could pursue Microfinance activities. Microfinance can be employed by organizations of different sizes, including MNCs.

Considering the activities undertaken, a distinction between the variations is applicable. Hence, as discussed, narrow Microfinance is presumed to engage solely in Microcredit provision services without the provision of additional support while broad Microfinance provides a broader range of services, accompanied by extensive training and support (Karlan and Valdivia, 2011). It is further presumed that the number of engaged initiatives is higher for the broad Microfinance approach than for the narrow strategy.

Underlying Business Notion

Generalizing the underlying business approaches of Microfinance is risky as they vary on a case-to-case basis. However, it can be said, that besides the societal motivation, the business motivation might be the positive impact on stakeholder relations for the narrow approach, and the potential for a profitable business case for the broad approach. Hence, business benefits can be either financial, in terms of revenue generation and growth opportunities, or non-financial, in terms of reputational benefits and reduced stakeholder pressure, contingent on the approach employed (Shetty, 2010). In line with this reasoning is the matter of business scalability which implies that the more economically sustainable the Microfinance operation, the more likely its replication on a broader scale. Accordingly, self-sufficiency is presumed to be larger for the broad Microfinance variation than for the narrow approach.

Compatibility

Microfinance is highly compatible with other business approaches on poverty alleviation. Concerning the remaining business strategies discussed, its compliance with CP is very likely if Microfinance is executed in a charitable fashion (Hossain and Knight, 2008). Moreover, the combination with inclusive BOP is possible, due to the shared idea of the poor as potentially talented entrepreneurs (Pitta, Guesalaga and Marshall, 2008). Lastly, Microfinance could also be undertaken as a social business if social goals would be the central objective of the business and financial motives would solely be secondary. Furthermore, Microfinance could be employed in a partnering approach with other societal actors, such as local NGOs or donor agencies (Rowe et al., 2009). Potential advantages of such cooperation could be foremost gained local knowledge as well as ease in approaching the poor due to the solid networks of local NGOs. Generally, the combination of Microfinance with other poverty-related approaches is highly preferable as it appears to increase its effectiveness on tackling poverty. To put it in the words of Hossain and Knight (2008, p. 167): "Although poverty cannot be single-handedly alleviated by Microfinance, it does play a large role in poverty reduction."

Impact on Poverty Alleviation

The narrow Microfinance approach views poverty primarily in economic terms, thereby generally pursuing an indirect poverty reduction strategy. In contrast, the broad strategy focuses, through additional provision of support and training, on both economic and non-economic poverty dimensions (Shetty, 2010). Initiatives are built in a bottom-up manner and active involvement of the poor as entrepreneurs is essential to "unleash people's potential" (Yunus and Jolin, 1999, p. 172). To be concrete, these strategies view the lack of financing possibilities as a root cause for poverty and thus aim at addressing the manner by providing possibility for financial access to the poor (Yunus and Jolin, 1999). Enhanced monetary access would in turn lead to reduced vulnerability and an ability to transform the money obtained into a stream of income by pursuing small-scale entrepreneurial activities, financed by small start-up loans (Shetty, 2010). The belief hereunder is that the obtained financial gains through entrepreneurship would be substantial enough to permanently move the poor beyond the poverty line.

Evaluation of Strategy

In practice, the results of Microfinance are mixed and range from highly positive to negative effects on poverty levels (Eder and Öz, 2008; Johnson, 2009). Again, contingency on implementation means and contextual factors has to be highlighted. On average, it can be concluded that Microfinance appears to positively influence poverty conditions and reduce vulnerability, but these gains are generally insufficient to permanently lift people out of poverty (Makita, 2011; Shetty, 2010). A possible explanation therefore is the lack of entrepreneurial knowledge which restricts the opportunity to operate a viable business (Karher, Iyani and Shannon, 2007). The broad approach to Microfinance thus aims at addressing this shortcoming by providing additional business and skills training to the poor to enhance their likelihood of success. In turn, changes of business success are slightly higher under the broad Microfinance approach and hence the positive effects of poverty reduction are more substantial. In addition, favorable effects in terms of advanced business knowledge and skills enhancement have been reported (Karlan and Valdivia, 2011).

Nonetheless – be it the narrow or the broad Microfinance approach – additional shortcomings must be mentioned. First of all, it has been said that certain Microfinance institutions foremost focus is on the moderately poor. In contrast, the poorest of the poor are continuously neglected due to their marginal credit worthiness which may in turn worsen social exclusion (Wright and Dondo, 2001; Rowe et al., 2009). In addition, it has been criticized that loans are not solely provided for entrepreneurial start-up activities but also for numerous consumer goods, which might not be essential for survival but would improve their quality of life. Along these lines, several incidents have been reported were access to credit has actually worsened the poverty situations of some due to inabilities to repay loans on time (Pitta, Gueslaga and Marshall, 2008). Furthermore, criticisms have accumulated about the terms and conditions of loan provisions, such as the charge of unreasonably high interest rates (Karher, Iyani and Shannon, 2007). Lastly, a final essential shortcoming is the fact that entrepreneurial activities are generally considered less stable and risky while the number of viable business opportunities is restricted (Karher, Iyani and Shannon, 2007). Hence, a more viable strategy for poverty reduction would be to increase the level of regular employment for long-term, stable income generation (Karnani, 2007a).

The Classification

Based on the previous analysis, the narrow Microfinance approach has been classified as *reactive poverty alleviation approach* while broad Microfinance initiatives are defined as an *active poverty alleviation approach.* The ultimate classification is visualized in Figure 2-10 and a detailed classification table of both Microfinance variations with further information can be obtained in Table 5-18 of Appendix E. To summarize, broad Microfinance appears to be more effective and sustainable in poverty alleviation than narrow Microfinance due to the additional skills and capacity building trainings provided as well as the financial self-sufficiency. However, both approaches solely address one small aspect of poverty, namely the lack of financial access, and neglect remaining non-financial poverty dimensions. Moreover, due to the fact that viable employment opportunities are limited, the magnitude by which Microfinance can permanently lift people out of poverty is also restricted.

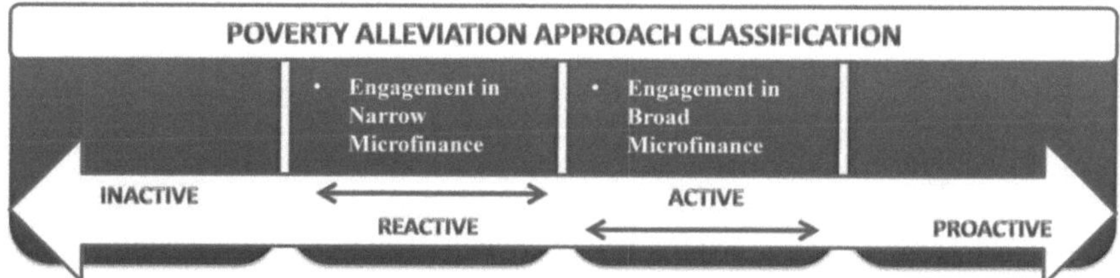

Figure 2-10: Classification of Microfinance

2.3.5 Bottom of the Pyramid

This subsection discusses the Bottom-of the-Pyramid strategy and its impact of poverty reduction.

The Approach and its Variations

Bottom of the Pyramid (BOP) is a business strategy on poverty alleviation which has been introduced by Prahalad and Hart in 2002. The main notion of that concept is illustrated with help of the worldwide economic pyramid – as visualized in Figure 2-11 – which displays the average purchase parity per household. Prahalad and Hart (2002) argue that most firms currently focus on the top tiers of the economic pyramid. Since these markets are already highly saturated, future growth potential will be limited. Instead companies should adapt their business models in order to address the bottom tier of the economic pyramid (as marked in red in the Figure 2-11) and thereby target the poor as customers (Prahalad and Hart, 2002).

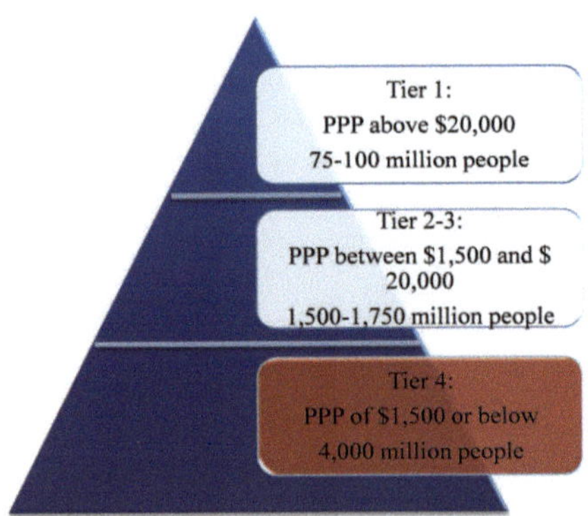

Figure 2-11: The Economic Pyramid of the World in terms of the Yearly Purchase Power Parity (PPP) per Household (Based on Prahalad and Hart, 2002)

More concretely, it has been argued that while the individual purchasing power of the BOP consumers may be marginal, the aggregate purchasing power of the 4 billion poor is substantial (Pitta, Guesalaga and Marshall, 2008). By providing value-adding products to the poor, their quality of life can be noticeably enhanced which will assist them in combating poverty (Olsen and Boxenbaum, 2009). Hence, Prahalad and Hart regard company engagement at the BOP as a win-win situation where firms can make a profit while simultaneously assisting in alleviating poverty (Prahalad and Hart, 2002).

Along these lines it has to be remarked that certain researchers refer to the *Bottom*-of-the-Pyramid, whereas others use the term *Base*-of-the-Pyramid. The general distinction among the two is the fact that the Base-of-the-Pyramid focuses on a broader group, namely the moderately poor which live of less than $2 a day whereas the Bottom-of-the-Pyramid tends to precisely concentrate on the poorest of the poor: the extremely poor who have to survive with less than $1.25 per day (van den Waeyenberg, 2011). Accordingly, due to the, in comparison, larger purchasing power of the moderately poor, Prahalad and Hammond (2002, p. 57) particularly describe the Base-of-the-Pyramid notion as "...a new market-driven paradigm for addressing poverty'. However, to facilitate the discussion, the proceeding analysis refers to both strategies in general as Bottom-of-the-Pyramid (BOP) while keeping the distinct interpretations in mind in further analyses.

With regard to the variations, the notion(s) previously discussed can be defined as *narrow BOP approach,* where the poor are primarily viewed as consumers. This view has been extended by a broader, more *inclusive BOP approach* which regards the poor not only as consumers but also as skilled producers, employees and entrepreneurs (Karnani, 2007b; Bais, 2008). This view was popularized under the lead of Karnani (2007b) who believes that the poor can solely be lifted out of poverty via long-term employment generation and inclusion into companies' value chains.

The Research Field

In comparison to most other discussed strategies, BOP is a relatively well researched area. Despite slight inconsistencies, findings are also quite homogeneous (Pitta, Guesalaga and Marshall, 2008). However, it becomes apparent that the majority of articles discuss BOP from a business perspective (van den Waeyenberg, 2011). In contrast, while the potential contribution of BOP to poverty alleviation is widely emphasized, a lack of research on the concrete implications of such undertakings for the poor can be obtained (London, 2007; Pitta, Guesalaga and Marshall, 2008). On the whole, it should also be noted that research on the narrow BOP approach is much more substantial although academia has recently began to increasingly discuss the inclusive BOP notion as well (London, 2007). An overview of the most crucial literature review citations on the impact of the BOP on poverty alleviation is revealed in Table 5-11 of Appendix C.

Organizational Characteristics and Activities

With regard to company size, BOP has classically been aimed at MNCs and this notion still prevails (Prahalad and Hammond, 2002). Besides, no restrictions on employed sectors can be undertaken. BOP initiatives can be pursued twofold, namely as part of a firm's peripheral activities or as being at the core of a company's operations, with the latter being more likely for the inclusive BOP approach and the former being generally accompanied by the narrow BOP variation (Olsen and Boxenbaum, 2009).

Concerning the undertaken activities, narrow BOP tends to focus on the sale of goods and services to the poor whereas inclusive BOP aims at including the poor in a company's value chain at various places (Prahalad and Hart, 2002; Karnani, 2007b). Once again, activities are presumed to be more extensive for the inclusive BOP strategy than for the narrow BOP approach.

Underlying Business Notion

The dominant business motivation for an engagement in narrow and inclusive BOP approaches is the business case at hand, and thus the potential for revenue generation and business growth (Prahalad and Hammond, 2002). Moreover, the inclusive BOP approach could further benefit from gains through skilled employees and favorable value chain reconfigurations via the involvement of the poor (Bais, 2008). Hence, benefits obtained could be in financial and non-financial terms. In either variation the financial sustainability of the BOP approach is high which in turn makes scalability likely – as long as profitable business opportunities can be obtained (Pitta, Guesalaga and Marshall, 2008). This business case approach is also the reason why BOP has frequently been praised as a benefitting win-win approach for business and society.

Compatibility

The compatibility of the BOP variations with other approaches is generally high even though it is expected to be higher for the inclusive notion than for the narrow BOP strategy. Engagement in partnerships with governments, NGOs and the poor is likely in order to access local knowledge and to obtain networking possibilities (Prahalad and Hammond, 2002; Bais, 2008). Additionally, the combination with other business strategies on poverty alleviation is possible as well, in particular with SE, CP and Microfinance (McKague et al., 2004). Once again, such a joint approach is presumed to increase the effectiveness of operations.

Impact on Poverty Alleviation

All in all, the potential contribution of the two BOP variations on poverty reduction seems to be substantial, even though the positive impact of the inclusive approach appears to be superior to the narrow initiative (Karnani, 2007b). Under both approaches, poverty is conceptualized foremost in an indirect, bottom-up matter with both economic and social poverty dimensions being addressed (Bais, 2008). Inclusion and empowerment of the poor is a vital part of the strategies, be it whether the poor are viewed as consumers (narrow approach) or as valuable employees, entrepreneurs or producers (inclusive approach) (Viswanathan et al., 2009). The benefits obtained by the poor are vast but varying. They can include financial gains in form of more favorable employment opportunities – mainly through the inclusive variation – as well as non-financial benefits, such as enhanced quality of life through better addressed needs or social inclusion and capacity building (Walsh, 2005).

Evaluation of Strategy

Although the potential contribution of the BOP is widely discussed, so are the negative aspects of the strategy, in particular with regard to the narrow variation. Hereby, criticism relates to the fact that simply addressing neglected consumer needs is insufficient to permanently lift the poor out of poverty (Karnani, 2007b). Along these lines, it should be stressed that the poor may as a consequence make questionable product choices and spend their money on goods, which they cannot afford and thereby fall deeper into debt. Instead, the provision of sustainable job opportunities for greater income has to be addressed. Another point of criticism encompasses the lack of inclusion of the poorest of the poor under both variations due to an emphasis on the base – and not the bottom – of the pyramid to obtain higher profit margins (Kelley, Werhane and Hartman, 2008). Besides, the BOP notion is criticized for its possibilities to crowd out local firms. Moreover, the widespread applicability of the BOP strategies are restricted by the extent to which profitable business opportunities in the BOP can be found, which according to several research studies, is not as high as previously claimed (Kelley, Werhane and Hartman, 2008). Nonetheless, besides the outlined critique, the overall effectiveness of the inclusive BOP strategy is especially high – if employed in the correct manner.

The Classification

The prior investigation has resulted in a classification of the narrow BOP variation as an *active poverty alleviation approach* and the inclusive BOP strategy as a *proactive poverty alleviation approach*. The resulting classification is displayed in Figure 2-12 and an extensive summary on the impact of the BOP notion on poverty can be penetrated in Table 5-19 of Appendix E. This classification led back to the fact that self-sustaining win-win business strategies appear to have the highest potential of ending poverty, under the condition that the impact on the poor is sustainable. Due to the fact that the long-term focus is higher for the inclusive approach, this strategy has been classified as more favorable.

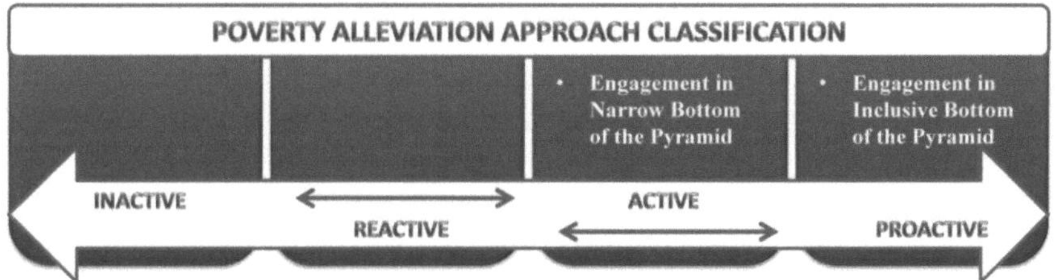

Figure 2-12: Classification of Bottom of the Pyramid

2.3.6 Social Entrepreneurship

This subsection will analyze the strategy of Social Entrepreneurship and investigate its poverty-alleviating potential.

The Approach and its Variations

The final presented business strategy on poverty alleviation is *Social Entrepreneurship* (SE). A wide variety of definitions for SE exists and numerous terms are used for the same phenomenon, such as social entrepreneurship, social business, social venture or social venturing (Yunus, 2010; Easterly and Miesing, 2009). In this regard, for purpose of simplicity, all these terms will be encompassed within the concept of *SE* for now, with the operating entity being called *social business*. Overall, most definitions share a common feature, namely that they classify a social business as a business entity which focuses primarily on social aims while financial goals are secondary in nature (Austin, Stevenson and Wei-Skillern, 2006). Hence, the SE will be broadly defined as "...entrepreneurships that produces products and services that directly cater to social needs" (Seelos and Mair, 2004, p. 4). Even though all social business strategies emphasize social goals over financial goals, they still differ in respect to the extent in which they do so. In this regard, the classification in Table 2-6, based on Peredo and McLean's (2005), has been proposed to in order to provide a better overview of SE conceptualizations.

Goals of Social Entreprise	Role of Commercial exchange
Entirely social	No commercial exchange
Entirely social	Some commercial exchange; profits benefit or support enterprise
Mainly social, partially financial	Commercial exchange; profits partly benefit entrepreneur and supporters
Social among others, including financial	Commercial exchange; profits mainly benefit entrepreneur and supporters
Subordinately social, primarily financial	Commercial exchange; profits almost entirely benefit entrepreneur and supporters

Table 2-6: Conceptions of Social Entrepreneurship Approaches (based on Peredo and McLean, 2005)

SE can employ many forms and address distinct issues, such as poverty. Two concrete variations should thus be distinguished. First of all, a *narrow SE notion* refers to operations targeted at poverty alleviation but in a non-financially-sustainable manner. And second, a *broad SE approach* implies that poverty reduction initiatives are undertaken under a financially self-sufficient configuration (Austin, Stevenson and Wei-Skillern, 2006). A further distinction among the two variations is that the broad SE approach regards poverty alleviation as its main social goal whereas under the narrow strategy additional social goals may be considered as well.

The Research Field

SE is a weakly understood and ill-defined concept in academia due to the fact that conceptualizations diverge (Bhownick, 2011; Weerawardena and Mort, 2006). While several researchers have attempted at clarifying the conceptions, still no universally accepted SE definition can be found (Weerawardena and Mort, 2006) Existing research focuses foremost on conceptualizing SE or on defining the success characteristics of social entrepreneurs and social businesses. In contrast, the potential impact of SE on poverty alleviation is only vaguely studied with a grave, current lack of empirical evidence (Mair and Schoen, 2005; Mair and Marti, 2004). An overview of the most important literature review findings thus far on the impact of SE on poverty alleviation can be seen in Table 5-12 of Appendix C.

Organizational Characteristics and Activities

The organizational characteristics for SE are numerous. Companies which engage in SE can be of any size, including MNCs, and originate from virtually all business sectors. Usually, SE strategies will be regarded as part of a firm's core activities, especially when considering the broad SE strategy (Austin, Stevenson and Wei-Skillern, 2006). Concerning the activities undertaken, companies which pursue a broad SE strategy are expected to engage in own SE initiatives. In contrast, companies with a narrow strategy might simply support external social enterprises without actually establishing a social business themselves. Respectively, it is further presumed that the number of SE activities undertaken is more substantial for the broad version in comparison to the narrow strategy.

Underlying Business Approach

The business goals within the SE strategy are secondary due to the fact that this strategy foremost aims at obtaining social and not financial benefits (Seelos and Mair, 2005). If the potential financial gains are nonetheless considered, the possible benefits obtained would fall under enhanced stakeholder relations and the potential for sustainable societal value generation (McIntyre, Bruno and Guerra, 2008). A consideration of financial benefits is thought to be higher for the narrow SE variation. In general terms, the self-sufficiency and thus the scalability are high for the broad SE notion, while being moderate for the narrow SE variation.

Compatibility

Similarly, the potential for compatibility with other approaches is high as well, and such combinations are assumed to increase the effectiveness of operations. Partnering approaches with the poor, governments and NGOs are highly regarded due to the possible knowledge gains obtained (Easterly and Miesing, 2009; van Slyke and Newmann, 2006). Furthermore, a joint implementation of several business strategies is considered beneficial, whereby particularly BOP, SE and Microfinance are favored (Goldsmith, 2011).

Impact on Poverty Alleviation

All in all, the positive impact of SE on poverty alleviation appears to be substantial. However, SE strategies can be pursued in a wide variety of manners, focused on distinct poverty dimensions, which in turn challenge a generalized analysis (Weerawardena and Mort, 2006). In principle, the addressed poverty dimensions can be economic or non-economic and the poverty alleviation approach can be indirect or direct. However, initiatives are almost exclusively performed under a bottom-up perspective which is accompanied by high involvement and empowerment of the poor (Yunus, 2010). The poor are thus viewed as skilled individuals, which can be supported in order to become independent in the long run. Potential benefits obtained by the poor could therefore range from financial gains, such as employment opportunities and income generation, to non-financial advantages via empowerment, social inclusion, capacity building or more favorable environmental conditions (Easterly and Miesing, 2009; Smith and Feldman Barr, 2007; Martin and Osberg, 2007).

Evaluation of Strategy

Besides the numerous positive aspects highlighted in the discussion, a frequent shortcoming of the SE strategies is the enormous diversity of approaches, which complicates comparison and scalability efforts (Mair and Marti, 2004). Moreover, especially the narrow SE approach is highly restricted financially for replication and expansion due to the dependence on external funds. A restriction of feasible, self-sustaining social business cases is a further limitation, particularly of the broad strategy. Overall, sustainability and effectiveness of SE on poverty reduction is substantial, even though it is much higher for the broad approach than for the narrow variation (Easterly and Miesing, 2009).

The Classification

All in all, based on the previous examination, the narrow SE strategy can be defined as *active poverty alleviation approach* whereas the broad SE strategy classifies as *proactive poverty alleviation approach*. This classification is displayed in Figure 2-13 and a more detailed classification table of the SE variations can be retrieved from Table 5-20 of Appendix E. The distinction among SE approaches is based on the focus on poverty and on the requirements for external funds in the narrow approach, which restricts self-sufficiency and scalability of undertakings. In contrast, the broach approach is entirely self-sustaining and hence more scalable, more extensive and more focused on combating the issue of poverty.

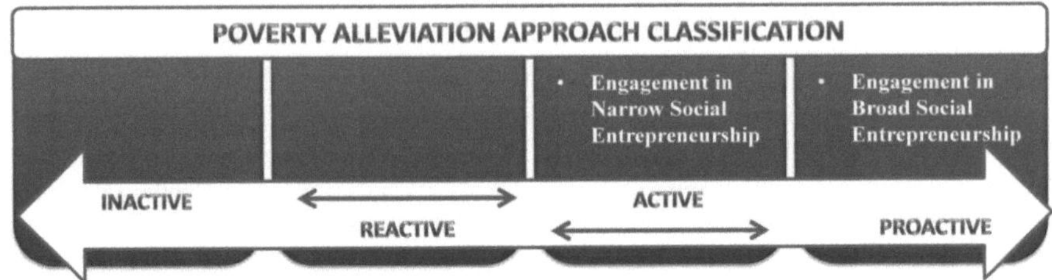

Figure 2-13: Classification of Social Entrepreneurship

2.3.7 Classification of Business Strategies on Poverty Alleviation

In this section five distinct business strategies on poverty alleviation have been analyzed. This subsection therefore aims at summarizing these classifications as well as the general findings obtained.

The Presented Business Strategies on Poverty Alleviation

The prior discussion has made clear that all discussed business strategies – with the exception of Business-as-Usual – consist of two distinct variations, namely a narrow and a broad/ inclusive approach. Figure 2-14 thus presents an overview of the examined business strategies and their variations.

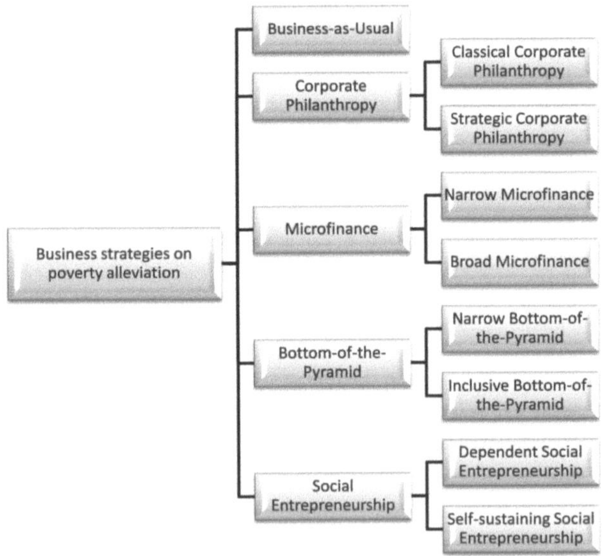

Figure 2-14: Overview of Analyzed Business Strategies on Poverty Alleviation and their Variations

The Classification

Figure 2-15 summarizes the resulting classification of the five business strategies and their variations based on the analysis in the prior subsections. On the whole, it can be seen that not all presented initiatives appear to be effective in sustainably tackling poverty. The majority of the investigated strategies fall into the active category. In general, Business-as-Usual is classified as least suitable approach whereas the inclusive BOP variation and the broad SE notion are ranked as most effective approaches.

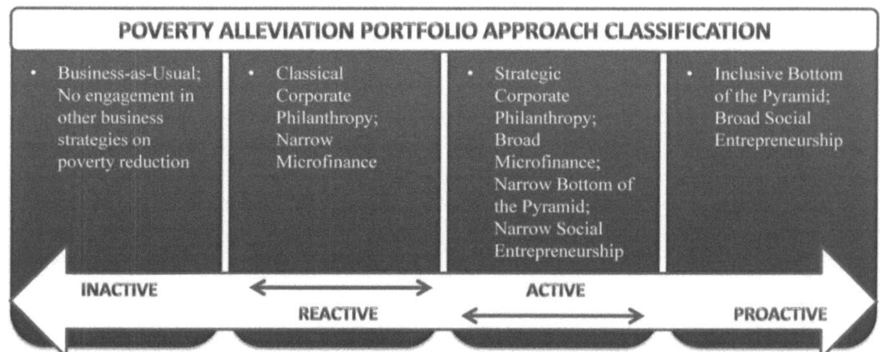

Figure 2-15: Classification of the Presented Business Strategies on Poverty Alleviation

A Concluding Remark

To summarize, this chapter defined numerous CSR-related terms and classified a variety of business strategies on poverty alleviation. This investigation revealed that the *how* – so the manner in which companies should best address poverty – is by no means straightforward. All in all, research findings are relatively subjective, vague and ambiguous which makes an objective comparison of the initiatives challenging. Nonetheless, the presented classification should enable a preliminary answer to the 'how question' by outlining which business strategies, and which variations of these strategies, might be more or less effective in tackling poverty. In this manner, this section provided an overview and comparison of initiatives at hand which should enrich the understanding of the research area.

2.4 Cross-Sector Partnerships and Poverty Alleviation Portfolios

This section discusses the theme of Cross-Sector Partnerships (CSPs) as well as the notion of Poverty Alleviation Portfolios. As a result, first the concept of CSPs and their potential contributions on poverty reduction is examined. This is followed by an analysis of the compatibility of the presented business approaches towards poverty alleviation with one another as well as the discussion on the configuration of Poverty Alleviation Portfolios. This section ends with an overall classification of all examined business strategies on poverty alleviation and a concluding remark on the entire theoretical part of this thesis.

2.4.1 Cross-Sector Partnerships and their Role in Poverty Alleviation

This subsection discusses CSPs and their potential impact on poverty reduction and is structured comparably to the sections – namely 2.3.2 until 2.3.6 – on the distinct business strategies.

The Approach

The term *Cross-Sector Partnership* (CSP) refers to partnerships between actors of the three distinct spheres in the societal triangle, being market, state and civil society. With regard to CSPs involving business, three main types can be distinguished, as visualized in Figure 2-16. The first type encompasses *Private-Nonprofit Partnerships* between firms and NGOs or poor communities which are undertaken primarily for the purpose of economic development initiatives and social issue addressing, such as the matter of poverty (Selsky and Parker, 2005). The second type of CSP is related to *Public-Private Partnerships* between companies and governmental actors primarily for the provision of public services or infrastructure-related causes (Selsky and Parker, 2005). Ultimately, the third common CSP type is the *Tripartite Partnership* between all three societal spheres, which aims at executing large-scale multi-sector projects. In this regard, a focused tripartite partnership approach needs to be highlighted, namely *Cross-Sector Social Oriented Partnerships* which concentrates entirely on tackling societal problems, such as poverty reduction (Selsky and Parker, 2005). To summarize, all three presented CSP types can address the issue of poverty but are most likely to do so in various manners.

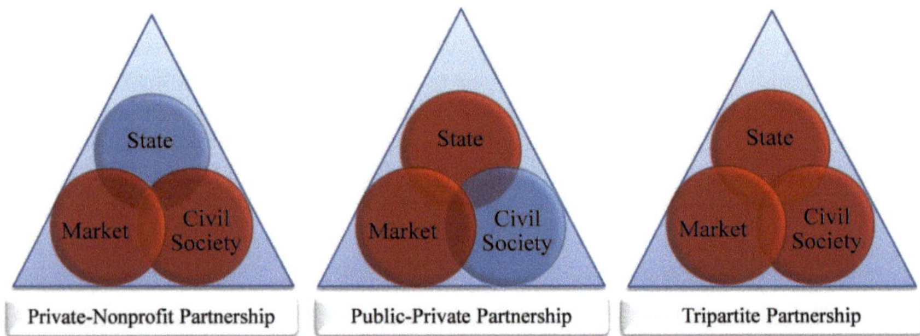

Figure 2-16: Types of Cross-Sector Partnerships Involving the Private Sector (based on Selsky and Parker, 2005)

The Research Field

Research among academics in the area of CSPs has greatly increased within the last years and the overall importance of partnerships for successfully tackling poverty has been frequently emphasized (Campbell, 2005). Still, the research field remains characterized by inconsistency and ambiguity (Brinkerhoff and Brinkerhoff, 2011). Moreover, analyses of key success factors for partnership configurations, as well as empirical studies on the concrete impact of CSPs on poverty alleviation are still lacking. In related terms, Kolk, van Tulder and Kostwinder (2008, p. 262) remark that: "... the potential contribution of companies as partners in furthering development objectives is frequently mentioned, but has received limited research attention." This highlights the need for more concrete investigation of the precise manner in which CSP can contribute to addressing the distinct poverty alleviation dimensions. An overview of literature review findings on the relevance of CSPs for poverty alleviation can be obtained in Table 5-13 of Appendix D.

Organizational Characteristics and Activities

Concerning the organizational characteristics of firms engaging in CSPs, no restrictions can be indicated. Firms can be of any size, including MNCs, and originate from any business sector (Arya and Salk, 2006). Generally speaking, CSPs are considered to be part of the core activities of a firm, but this is presumed to be more likely the more extensive a firm's partnering approach. The activities undertaken in a partnership are various and are highly contingent on the particular project with a more in-depth analysis of the activities provided in the 'partnering approach' paragraph below.

Underlying Business Notion

The underlying rationale for the emergence of CSP is the fact that solving certain societal issues, such as poverty, is beyond the ability of a single actor (Humphrey, 2006). Instead collaborative efforts between distinct actors are required in order to solve such matters. In this context, the following reasoning for engagement in CSPs can be employed: "The main advantage of collaborative agreements lies in the pooling of resources and knowledge leveraged by the partners, which may in turn lead to the development of a broader portfolio of resources for firms in the network." (Yunus, Moingeon and Lehmann-Ortega, 2010, p. 314). Hence, the underlying business notion of an engagement in CSPs lies in its potential of accessing complementary resources which would in turn increase the effectiveness of operations (Brinkerhoff and Brinkerhoff, 2011). Concrete benefits obtained include additional sources of local knowledge and expertise, in addition to enhanced opportunities for networking and further funding sources (Andrews and Entwistle, 2010). Moreover, engagement in responsible initiatives tends to positively impact stakeholder relations (Oketch, 2004). CSPs are usually accompanied by high financial self-sufficiency due to the large number of actors involved, which in turn is favorable for the scalability of such undertakings.

Partnering Approach

Partnering approaches can occur under a wide variety of configurations (Arya and Salk, 2006). As previously discussed, the common forms which include companies are Private-Nonprofit Partnerships, Public-Private Partnerships or Tripartite Partnerships. Hence, firms can corporate with NGOs, governmental institutions or poor communities either individually or at once. Besides the different types of CSPs, a variation also should be made among distinct approaches based on the extensiveness of the initiatives. Hereby, important distinguishing characteristics include the number of initiatives, the amount of partners, the scope and duration of the projects as well as the form of engagement (Wasmer, 2008; da Rosa, Barendse and Kazarjan, 2011). Research results on these distinct aspects are partially conflicting and case-dependent so solely certain general indications can be made. As an example, the importance of long-term partnerships has been stressed as superior in past research (Yunus, Moingeon and Lehmann-Ortega, 2010). Next, with regard to size, Arya and Salk (2006, p. 221) claim that "... up to a threshold level, the greater the number of partners that contribute in terms of open discussion on best practices, costs, risks and expertise related to socially responsible behavior, the greater the likelihood that corporations progress among the learning curve to becoming more socially responsible."

Compatibility with Business Strategies
As discussed in prior section, CSPs are highly compatible with all of the previously presented business strategies on poverty reduction, except for Business-as-Usual due to its low priority on poverty (Rowe et al., 2009; Newell and Frynas, 2007; Moran, 2008). Hereby, the quotes in Table 5-15 of Appendix D highlight that employing these initiatives via a partnering approach is beneficial for increasing the likelihood of success in combating poverty. Accordingly, it has been announced that "... cooperation is considered as a major factor of success for pro-active CSR strategies" (Yunus, Moingeon and Lehmann-Ortega, 2010, p. 314). In the same notion, it can be concluded that the more sustainable the business strategy, the greater the potential contribution to poverty.

Impact on Poverty Alleviation
The overall importance of CSPs for combating social problems, such as poverty alleviation, has been repeatedly stressed in past studies (Mert, 2009; Warhust, 2005; Humphrey, 2006). However, due to the fact that CSPs diverge largely in their orientation and configuration, it is challenging to generalize their impact on poverty elimination. Instead, it is more useful to say, that they can be highly effective in addressing poverty, if implemented in an appropriate manner (Rein and Stott, 2009). The strength of employing partnership approaches lays precisely in its versatility which implies that nearly all poverty dimensions can be targeted. Hence, it is presumed that the more poverty dimensions addressed, the greater the impact on poverty alleviation, which would in turn make a diverse portfolio favorable (Vurro, Dacin and Perrini, 2010). Moreover, CSPs are regarded as sustainable due to the usually high involvement and empowerment of the poor, which is crucial to the path towards independence (Lapeyre, 2011).

Evaluation of Strategy
All in all, CSPs appear to be highly effective approaches for tackling poverty. Nonetheless, critics argue that this effectiveness is contingent on the manner in which the approach is employed and thus no 'one-size-fits-all' approach can be presumed (Spielman, Hartwich and von Grebmer, 2007; Rein and Stott, 2009). Thus, contextual key success factors are vital to fully nurture the potential of CSP as a poverty alleviation tool. As a general rule it can be presumed that the more extensive the partnering approach, the greater the effectiveness in tackling poverty (Wasmer, 2008). An extensive portfolio possesses several characteristics, including a high number of undertaken initiatives with a wide variety of actors, a large number of addressed poverty alleviation dimensions as well as a long-term sustainable orientation (Arya and Salk, 2006).

The Classification
In classifying CSPs in terms of the effectiveness regarding poverty alleviation, it can be concluded that CSPs generally are highly effective initiatives for combating poverty. However, due to their large variations the concrete classification is contingent on the precise partnering

approach employed, as visualized in Figure 2-17. Along these lines, Olsen and Boxenbaum, (2009) emphasize that collaboration has to be proactive in order to be effective.

Accordingly, based on the prior analysis, four distinct partnering approaches on poverty can be distinguished, including limited, moderate and extensive partnership approaches. First, a lack of engagement in any partnering activities implies that a company pursues *no partnering approach*. Second, a *limited partnership approach* denotes that a firm only superficially engages in partnering activities. Hence, the number of partnering activities is marginal, few partners and projects are sought in a limited number of areas, with projects that are humble in scope and duration. Moreover, no particular poverty focus of such partnerships can be observed. Third, a *moderate partnership approach* corresponds to a moderate number of partnering activities with several partners and projects. The scope and duration of these projects as well as the focus on poverty are modest. And fourth, an *extensive partnership approach* adheres to a substantial number of partnering initiatives, especially emphasizing poverty alleviation. Hence, partners are likely to be more plentiful and projects are expected to be more extensive in scope and duration. With regard to poverty alleviation it is thus presumed that the more extensive a partnership portfolio, the higher its possible contribution to poverty alleviation.

With regard to the classification of the approaches, the lack of a partnering approach on poverty reduction would be defined as an *inactive poverty alleviation strategy*. In turn, a reactive poverty alleviation approach would accord with a limited partnership, which encompasses a low number of undertaken initiatives with a narrow focus and a short-term orientation. Subsequently, an *active poverty alleviation approach* would correspond to a moderate partnering approach with a modest number of partnerships which are medium-term in scope. Lastly, a *proactive poverty alleviation approach* implies an extensive partnering portfolio with a substantial number of addressed poverty dimensions as well as initiatives which are high in number and long-term in scope. An overall analysis table of CSPs for further information can be found in Table 5-21 of Appendix E

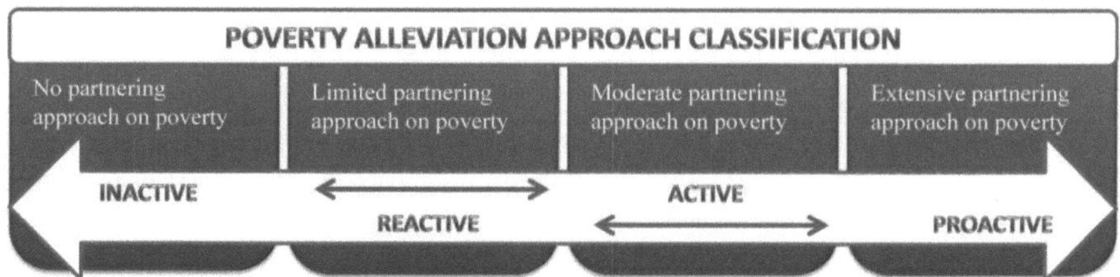

Figure 2-17: Classification of Cross-Sector Partnerships

2.4.2 Combinations of Poverty Alleviation Approaches

After outlining and classifying several distinct business strategies on poverty alleviation, as well as the concept of CSPs one starts wondering to which extent these approaches might be simultaneously combinable. A visual representation on the manner in which these potential initiatives could be combined is displayed in Figure 2-18. Hence, this subsection aims at summarizing existing literature on whether the stated business strategies can be employed in a partnership approach, and if so, whether such an undertaking would have positive implications for poverty reduction. Moreover, it is discussed whether the presented business strategies on poverty alleviation are simultaneously compatible. Ultimately, the notion of companies' Poverty Alleviation Portfolios is presented.

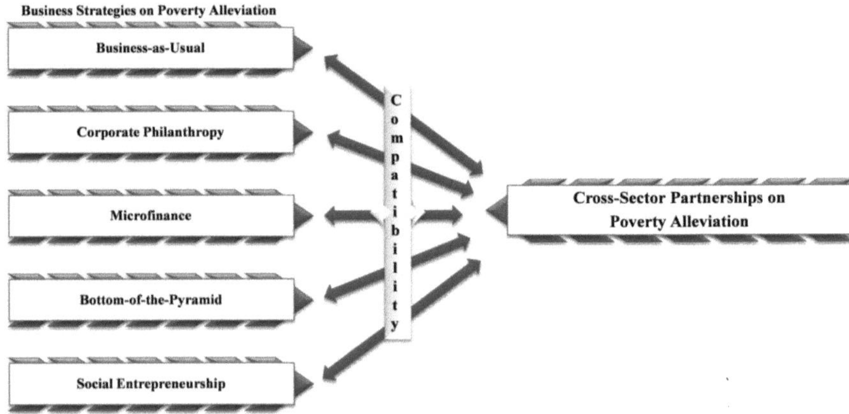

Figure 2-18: Combination Possibilities between Examined Poverty Alleviation Approaches

Employing Business Strategies in Partnership

As discussed in the previous section, all of the described business strategies on poverty alleviation can in theory be employed through a partnering approach. However, as pursuing a Business-as-Usual approach implies no focus on poverty-related initiatives, it is unlikely that such a strategy would be undertaken via a partnering approach which emphasizes the goal of poverty reduction (Warhust, 2005). In contrast, for the remaining four strategies as well as their variations, implementation through a partnership approach should be possible due to the presumed goal alignment in terms of addressing poverty (Tang and Li, 2009; McKague et al., 2004; Goldsmith, 2011; Moran, 2008; Rowe et al., 2009). In particular with regard to the more active and proactive initiatives, such as BOP and SE, the added value of employing a partnering approach with other societal actors has frequently been highlighted (Walsh, 2005). Hence, it appears to be recommendable to undertake business initiatives in partnership with other societal actors, as it is presumed to increase the effectiveness of the approaches in terms of combating poverty (Gardetti, 2005). Potential advantages obtained thereby are for instance the combination of complementary skills and resources, reduced operational costs as well as local knowledge and networking opportunities (Gifford, Kestler and Anand, 2010; Pitta, Guesalaga and Marshall, 2008). However, it also becomes apparent that such joint approaches – even though extremely favorable – are a highly under researched field as no concrete academic findings were presented on the precise manner in which such an undertaking could positively contribute to ending poverty. Still, from the provided indications it is presumed that firms' initiatives on poverty alleviation are more effective the higher the number of approaches and the more proactive the combination of approaches. An overview of literature review citations on the combination possibilities of the five business strategies with CSPs can be found in Table 5-14 of Appendix D.

Employing Business Strategies Jointly

Moreover, as has been examined in the previous sections, undertaking the presented business strategies on poverty reduction jointly – which implies employing two or more strategies simultaneously within a single project – seems to be highly beneficial in order to increase the effectiveness in eliminating poverty (Bais, 2008; Easterly and Miesing, 2009; van Slyke and Newmann, 2006; Hossain and Knight, 2008). This is due to the fact that different initiatives emphasize varying dimensions and accordingly address poverty in a distinct, complementary manner. Considering the extent to which the distinct strategies discussed above could be combined, a great variety of combinations is possible, as demonstrated in Table 2-7. More concretely, with the exception of Business-as-Usual, all possible combinations are theoretically feasible (Walsh, 2005). Employing a comparable reasoning to the one used above, Business-as-Usual is unlikely to be combined with the other approaches due to the fact that this strategy does not directly address poverty, while the others aim at doing so (Ayamwu, 2006). Hence, a joint undertaking would be a contradiction in itself. While certainly some initiatives are more effective in combating poverty than others it can nonetheless be presumed, based on the literature findings, that a greater variety of actual approaches would make a contribution to poverty reduction more likely (Goldsmith, 2011; Walsh, Kress and Beyerchen, 2005). As a consequence, it is expected that the greater the number of simultaneously pursued business strategies and the more sustainable the initiatives, the higher the potential impact on poverty alleviation. Nonetheless, it should once again be remarked that the literature on the combination of the business strategies is at present highly restricted and would benefit from further research, in terms of how such combinations could contribute to poverty alleviation. An overview of obtained literature review quotes on the simultaneous combination possibilities of the five business strategies can be found in Table 5-15 of Appendix D.

Approach	Business-as-Usual	CP	Microfinance	BOP	SE
Business-as-Usual	X	Low	Low	Low	Low
Corporate Philanthropy	X	X	High	High	High
Microfinance	X	X	X	High	High
Bottom of the Pyramid	X	X	X	X	High
Social Entrepreneurship	X	X	X	X	X

Table 2-7: Likelihood of Compliance of Business Strategies on Poverty Alleviation

Company Portfolios on Poverty Alleviation

The previous investigation has highlighted the increased importance of undertaking numerous business initiatives on poverty reduction at once. As a result, it should be beneficial to jointly analyze and compare a firms' Poverty Alleviation Portfolios. This is a novel term and represents a new unit of analysis in the field of business and poverty. In general a portfolio refers to the whole of activities in a certain respect undertaken by a single firm and, in turn, a *Poverty Alleviation Portfolio* would encompass all poverty-related activities pursued by a single firm (da Rosa, Barendse and Kazarjan, 2011).

The literature review undertaken in this chapter has not revealed a single prior examination of poverty-related firm portfolios. The term 'portfolio' has solely been employed by Leisinger (2007) in related matters when discussing firms portfolios of distinct, voluntary CSR initiatives. Instead the unit of analysis in past research has generally been the examined strategy and not the firm itself. In the context of CSPs, da Rosa, Barendse and Kazarjan (2011) obtained a similar finding by outlining that 'portfolio' is a novel term in CSP research. Besides the lack of companies' portfolio as unit of analysis, a/an (joint) exploration of the distinct variations and combination possibilities of initiatives on poverty has also been widely lacking in prior research.

Hence, this thesis employs distinct approaches than prior literature by examining firms' entire portfolios on poverty alleviation initiatives, in order to ultimately classify the effectiveness of such a portfolio approach in addressing poverty. The previous literature findings suggest that the more (proactive) initiatives a firm undertakes and the more engaged it is with regard to poverty-alleviating approaches, the higher will be its potential contribution to reducing poverty. Even though initiatives differ, as revealed above, in their effectiveness, it can hence nonetheless be presumed that an extensive firm portfolio on poverty is superior to a moderate, limited or even none existent portfolio with regard to its poverty-alleviating potential.

All in all, the prior discussion has revealed that the combination of distinct business approaches on poverty alleviation – be it within one project or on the whole – is highly advisable for tackling poverty. As a result, a classification and analysis of firms' entire portfolios of poverty-related initiatives would be beneficial in order to access the extent to which a company effectively addresses the issue of poverty.

Classification of the Combination of Poverty Alleviation Approaches

Based on the analysis in the prior subsections, companies' Poverty Alleviation Portfolio Approaches can be classified as revealed in Figure 2-19. Accordingly, a lack of combined approaches and no presence of a Poverty Alleviation Portfolio correspond to an *inactive approach.* Next, a low amount of undertaken initiatives, pursued predominantly in isolation indicates a limited Poverty Alleviation Portfolio and thus be categorized as a *reactive approach.* Besides, a moderate number of undertaken business initiatives that are foremost employed in combination is defined as a moderate Poverty Alleviation Portfolio and in turn ranks as an *active approach.* Lastly, a substantial number of pursued initiatives which are undertaken solely in combination correlates to an extensive Poverty Alleviation Portfolio and would hence be regarded as a *proactive poverty reduction approach.*

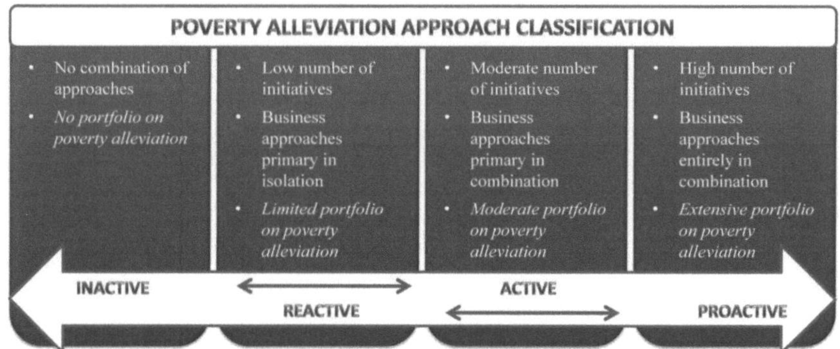

Figure 2-19: Classification of the Combination of Business Approaches on Poverty Alleviation

2.4.3 Classification of Companies' Poverty Alleviation Portfolio Approaches

Based on the prior discussions in section 2.3 and the remainder of 2.4, the presented business approaches towards poverty alleviation are jointly classified in this subsection, according to the employed CSR framework developed by van Tulder and van der Zwart (2006). In this regard, building upon all prior sub-classifications, the whole classification of firms' Poverty Alleviation Portfolio is comprised. Ultimately, the findings are discussed and a conclusion from the entire subsection as well as the complete theoretical literature review is drawn.

The Overall Classification

The previous results on the classification of poverty alleviation approaches as well as their variations and combinations are presented in Figure 2-20. Within this overarching classification several distinct aspects were evaluated – as indicated by the gridlines in the figure – which will now be summarized and unified in a single classification framework.

First of all, five business strategies on poverty reduction were evaluated. In this regard, solely Business-as-Usual classifies as inactive. In contrast, classical Corporate Philanthropy and narrow Microfinance are considered reactive approaches whereas strategic Corporate Philanthropy; broad Microfinance; narrow Bottom of the Pyramid and narrow Social Entrepreneurship are defined as active. Finally, inclusive Bottom of the Pyramid and broad Social Entrepreneurship are regarded as proactive initiatives.

Next, Cross-Sector Partnership approaches and their impact on poverty elimination were reviewed. Hereby, a lack of a partnering approach on poverty reduction is classified as inactive; a limited partnering approach as reactive; a moderate approach as active and an extensive approach as proactive.

Afterwards, the combination of approaches was investigated. In this regard, a lack of joint initiatives implies an inactive classification whereas a low number of approaches being pursues primarily in isolation is defined as reactive. In turn, a moderate number of undertaken initiatives that are implemented predominately in combination fall into an active approach while a high number of employed approaches which are entirely executed in a combined manner are considered a proactive poverty alleviation approach.

Thus, taken as a whole, an *inactive approach* on poverty alleviation implies the lack of a poverty-related portfolio; a *reactive approach* a limited portfolio; an *active approach* a moderate portfolio and ultimately a *proactive approach* an extensive portfolio of company initiatives on poverty reduction.

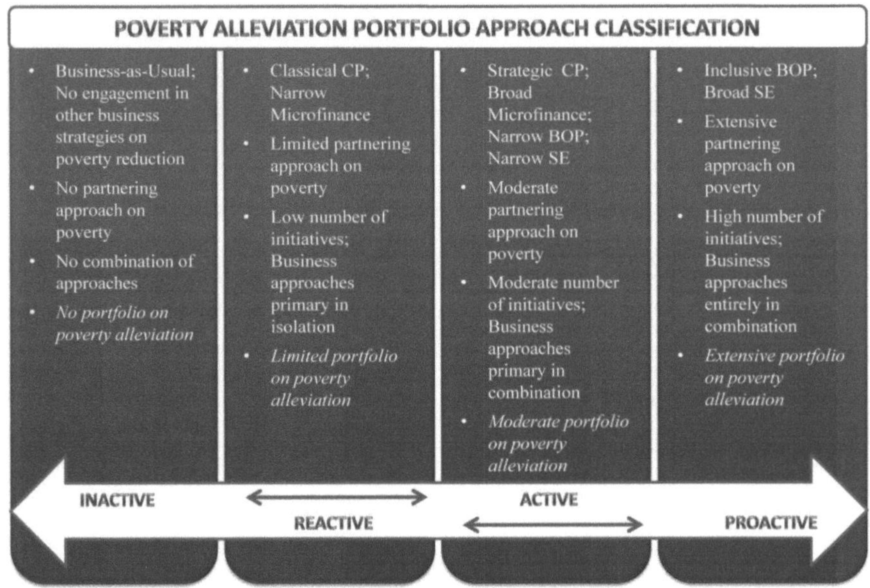

Figure 2-20: Classification of Companies' Poverty Alleviation Portfolio Approaches

Relating Theory to Practice

The theoretical literature review part of this thesis served as a revision of the discussion of business and poverty alleviation in academia. Overall, the goal hereby was to outline the general role which MNCs are expected to play with regard to poverty reduction as well as the possible business approaches they could thereby employ. Hence, this part aimed at addressing the '*whether*' and the '*how*' questions: Should companies engage in the field of poverty elimination? And if so, which approaches appear to be most promising in their societal impact?

Overall, it can be concluded that the literature on poverty is plentiful. However, in particular with regard to the concrete business roles and approaches, widespread controversies and ambiguities among researchers can be found. Many terms and approaches have a multitude of meanings or varieties and are thus vaguely defined. In addition, concerning the potential impact of business initiatives on poverty reduction, concrete empirical impact analyses are generally lacking and the potential implications are primarily presumed based on academic reasoning but not actually tested in practice. This underscores the need for empirical research on the actual engagement of companies with regard to poverty, as undertaken in this study. Hereby, the classification model developed in this theoretical part will be applied to practice.

On the one hand, the presented literature review findings are beneficial as they indicate that clearly a literature gap and a need for clarification can be recognized. But on the other hand, such a lack of clear and homogeneous findings in turn complicates an objective classification of approaches. Hence, it has to be kept in mind that a slight touch of subjectivity may have contributed to the present classification, even though it was primarily based on the literature review findings and analytical reasoning.

A Concluding Remark

Overall, the literature review section of this thesis aimed at examining the following four previously presented research questions:

- *What are the roles and potential contributions of the business sector, in general, and MNCs, in particular, towards combating poverty?*
- *In how far are the examined business strategies on poverty alleviation – being Business-as-Usual; Corporate Philanthropy; Microfinance; Bottom of the Pyramid and Social Entrepreneurship – and their variations effective in addressing poverty?*
- *In how far are Cross-Sector Partnerships effective in addressing poverty?*
- *In how far can the prevailing poverty alleviation initiatives be combined and in how far would such a joint approach increase the effectiveness of the undertakings?*

In this regard, the following conclusions can be made. Regarding the first research question, it can be concluded *that* the business sector overall as well as MNCs in particularly are widely recognized as playing a vital role in alleviating poverty. However, substantial ambiguity prevails among academia with regard to the manner of *how* business can and should assist in poverty reduction.

The second research question therefore concerned the discussion of several prevailing business strategies to poverty elimination. This analysis has revealed that the extent to which the presented initiatives are effective in poverty alleviation diverges. Accordingly, Business-as-Usual seems to be less effective in combating poverty, followed by Corporate Philanthropy and Microfinance. Hence, Bottom of the Pyramid and Social Entrepreneurship appear to be the most effective poverty reduction strategies. However, differences in the variations of the strategies also need to be considered.

The discussion of the third research question revealed that Cross-Sector Partnerships overall appear to be a highly effective approach with regard to ending poverty. However, outcomes are contingent on the partnering approach and the concrete method of implementation with an extensive partnering approach considered most effective, followed by moderate, limited and ultimately no partnering approaches.

With respect to the fourth research question, the structured literature review has revealed that all discussed business strategies on poverty alleviation, with the exception of Business-as-Usual, are likely to be employed in a partnering approach. Moreover, an overall combination of all discussed business strategies, besides Business-as-Usual, seems to be possible as well. Even though, empirical findings are sparse, results thus far indicate that a greater number of employed approaches as well as the usage of partnerships within the business strategies appear to increase the effectiveness in tackling poverty. Hence, it can be concluded that the more extensive and the more sustainable a firm's portfolio of poverty alleviation, the greater the potential impact on poverty alleviation.

3 The Primary Research on Multi-National Corporations

This chapter presents the primary research of this thesis. It consists of three sections and is structured as follows. In the first section the research methodology is described. In section two the results are presented. Finally, section three discusses the results and explores possible explanatory factors for the findings.

3.1 Methodology

In this section, the methodology of the conducted primary research is displayed. More precisely, the reasoning for the selected research approach and research design is briefly illustrated. Moreover, the data collection, codification and classification method as well as the sample design and the conceptual model are presented. Ultimately, the data analysis approach, the exploratory analysis and the research characteristics are described.

3.1.1 Research Approach

The literature review on business and poverty has revealed that the overwhelming majority of practical research in the field is qualitative in nature and based on a small number of descriptive case studies. Moreover, past research almost solely looks at one business strategy on poverty alleviation in isolation and Cross-Sector Partnerships are reviewed separately from the business strategies. No distinct variations or combinations of approaches are explored and the term portfolio in the context of poverty alleviation approaches is uncommon.

In turn, this research aims to address the gap between theoretical roles of companies regarding poverty alleviation and their actual contributions in practice by examining how extensive the *Poverty Alleviation Portfolios* of MNCs actually are. In doing so it directly applies the classification framework developed in the theoretical part. This study consequently takes a unique approach by jointly studying the business approaches undertaken by Fortune Global 100 firms on poverty alleviation. This should enrich existing literature due to the larger sample size. Furthermore, several previously separated fields of research are examined in conjunction in order to present a better overview of the entire Poverty Alleviation Portfolio per company. Hereby, initiatives with regard to the five discussed business strategies on poverty alleviation are reviewed at once. In addition to that, the topics of business strategies on poverty alleviation and the concept of Cross-Sector Partnerships are combined. Moreover, the joint combination of approaches is taken into account, which has been highly neglected in prior researched. By doing so, this approach should therefore permit a broad overview of the actual initiatives undertaken on combating poverty by the Fortune Global 100 firms, which consequently should enable a unique classification of these companies in terms of the Poverty Alleviation Portfolio Approaches. Such a classification could ultimately serve as a first evaluation on whether MNCs are engaging extensively in activities related to poverty reduction.

3.1.2 Research Design

Considering the research design, primary research data was collected and examined via *content analyses of companies' publicly available information* on poverty, in order to specifically address the research aim presented above (Solomon, Marshall and Stuart, 2008). More concretely, this primary research was conducted in the form of *qualitative research.* Accordingly, qualitative research implies that "... the research tends to be concerned with words rather than numbers [with emphasis being placed on] the understanding of the social world through an examination of the interpretation of that world" (Bryman and Bell, 2007, p. 402). One main research method which was employed in this case and which is strongly associated with qualitative research is "... the collection and qualitative analysis of texts and documents" (Bryman and Bell, 2007, p. 404). Furthermore, using publicly available documents as sources of data is also a common undertaking in qualitative research design (Bryman and Bell, 2007).

The research design is partially *descriptive* and partially *exploratory* in nature. Descriptive research is characterized as being observational by describing *what* can be observed whereas exploratory research analyzes *why* certain observations can be made by providing causal explanations (Solomon, Marshall and Stuart, 2008). In view of that, the obtained company information, with regard to firms' engagement in poverty alleviation approaches, is first described, then firms' Poverty Alleviation Portfolio Approaches are classified and ultimately possible underlying reasons for different classifications among companies are explored.

In order to facilitate interpretation of the data, the retrieved information was *codified* and *classified* into categories via content analyses – as explained in detail in latter parts (Bryman and Bell, 2007). A further remark at this point is that even though this research is on the whole qualitative in nature, a quantification of the obtained information was conducted to display the frequency in occurrence of the selected units of analysis. Bryman and Bell (2007) argue that a *quantification of data* for codification and classification purposes in a qualitative study is a valid and common research approach.

3.1.3 Data Collection Method

The primary data was collected through the means of content analyses on the self-*reporting of firms* with regard to poverty on their *company websites* and in their *CSR reports* (as obtained via the company websites). This data collection method is regarded as a valid research approach and has been pursued by several authors beforehand, including Helsin and Ochoa (2008); Wanderly et al. (2008); Tang and Li (2009); Chapple and Moon (2005); Belal and Cooper (2007); van Tulder (2010); or Fortanier and Kolk (2007). Accordingly, Tang and Li (2009, p. 204) state the following on CSR reporting: "Researchers have begun studying corporate social reporting as a proxy of companies' actual CSR practices and their PR strategies. Content analysis of companies' annual reports and CSR reports has been a frequently used method in the study of corporate social reporting in the social and environmental accounting literature since the 1970s."

Moreover, Wanderly et al. (2008, p. 370) remark in this context, that due to its rapid growth, "... the internet has become one of the main tools for CSR information disclosure." Overall, *online research* of corporate reporting is advantageous as a data collection method due to the fact that company websites are "...functionally uniform units of analysis" (Chapple and Moon, 2005, p. 424), with information being plentiful and easily accessible (Wanderly et al., 2008). In contrast, a main limitation of this data collection method is the fact that the information, which companies present publicly on their websites and CSP reports, can be selected according to the firm's interests. Hence, particular information can be excluded or modified in order to fit the desired firm image (Chapple and Moon, 2005). As a result, data validity cannot fully be guaranteed due to the subjectivity of the displayed information. However, past research has indicated that "... it is assumed here that, in general terms, the greater the extent of reporting, the more engaged the company is with CSR and the more seriously it is taken therein" (Chapple and Moon, 2005, p. 424). Thus, drawing conclusion from corporate self-reporting on actual firm behavior is a valid undertaking.

As briefly mentioned above, the obtained quotes from firms' reporting on their engagement with regard to poverty alleviation were collected, classified and studied through means of *content analyses*. Content analysis is a common research approach that has been employed in numerous previous studies, within the context of corporate CSR reporting, for instance by Wanderly et al. (2008) and Cochran and Wood (1984). In this regard, the following explanation of content analysis is provided, which perfectly suits the approach undertaken in this study:

"The second [generally accepted] method of measuring CSR is content analysis. Normally, in content analysis the extent of the reporting of CSR activities in various firm publications and especially in the annual report is measured. This can consist of simply noting whether or not a particular item is discussed either qualitatively or numerically, or it can mean actually counting a number of items. [...] Content analysis has two significant advantages. First, once the particular variable has been chosen (a subjective process), the procedure is reasonably objective. Therefore, the results are independent of the particular research. Second, because this technique is more mechanical, larger sample sizes are possible. However, content analysis also has some drawbacks. The choice of variables to measure is subjective. Further, content analysis is only an indication of what firms say they are doing, and this may be very different from what they are actually doing." (Cochran and Wood, 1984, p. 44).

3.1.4 Sample Design

In this study, emphasis was placed on *MNCs*, because of their enormous contributing potential to poverty alleviation. Moreover, focus on MNCs was also employed for practical reasons, due to the substantial amount of publicly available information that can generally be obtained for these firms. It was presumed that receiving information in similar levels of detail online would have beene more challenging for smaller firms which in turn could have hindered an accurate and elaborate analysis.

Concerning the concrete sampling, *systematic sampling* was chosen which implies the specification of a starting point and a fixed interval (Solomon, Marshall and Stuart, 2008). Due to the fact that MNCs are the target population of this research, a diverse but representative sample of large, international firms was favored. A relatively large *sample size of 100* should enable inference-making on the target population. In this regard, the largest 100 MNCs from the Fortune Global 500 firms list were selected, namely rank 1 to 100. The Global 100 list is constructed based on companies' yearly revenue and it was obtained from the Fortune website (2011). Firms from that list were chosen as a sample because of the fact that these firms possess the desirable characteristics of MNCs, being large in size and global in scope.

A concrete overview of the 100 sample firms can be found in Table 5-22 of Appendix F. This table contains a list of the selected companies in addition to presenting their country and region of origin, their company sector and their ultimate portfolio classifications. The country and in turn region of origin data was provided on the Fortune website as well (2011). In contrast, the sector was obtained according to a subsector classification from the Industry Qualification Benchmark website (2011). The region of the 100 firms is relatively diverse with 43 companies originating in Europe, 32 in North and South America and 25 in Asia. With regard to the company sector the similar rate of diversity prevails with 27 firms operating in the financial sector, 21 in oil and gas, 7 in telecommunications, 5 in utilities as well as 10 each respectively engaging in the automobile, the retail and the technology industry. Lastly, 10 firms were classified as 'others' to simplify the analysis, due to the fact that their sectorial classification was below frequency of five firms.

3.1.5 Conceptual Model

Based on the previously selected variables of analysis, a *conceptual model* has been developed which can be seen in Figure 3-2 below. However, in order to understand this model a revised figure of Figure 2-18 from the theoretical part is presented (see Figure 3-1). This figure displays the components which are examined and included in the ultimate classification. Hence four of the five discussed business strategies on poverty alleviation are classified, namely Corporate Philanthropy, Microfinance, Bottom of the Pyramid and Social Entrepreneurship. Business as Usual has been excluded from the classification due to the fact that persuasion of Business-as-Usual implies non-engagement in any of the other initiatives and is thus already included indirectly in the classification. Moreover, the classification of the Cross-Sector Partnerships approach on poverty alleviation is examined as an additional component. Ultimately, two distinct combination possibilities are also included in the final classification, namely the simultaneous combination of the four discussed business strategies as well as the application of these four respective strategies in a partnership approach.

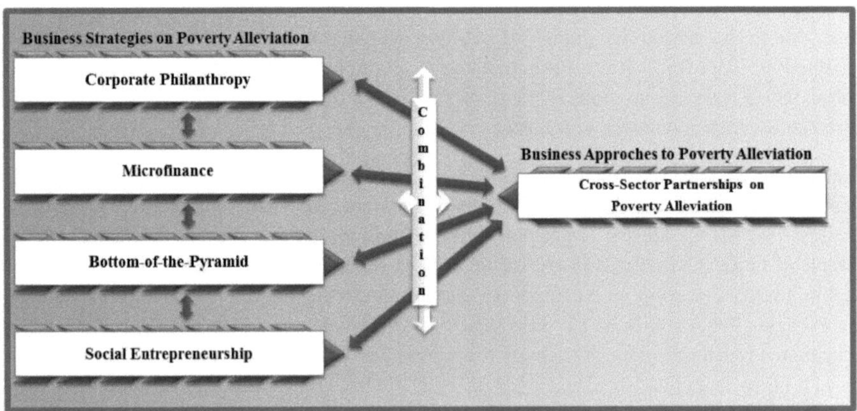

Figure 3-1: Examined Classification Components

As a result, the conceptual research model in Figure 3-2 describes how – based on all these distinct components – the ultimately Poverty Alleviation Portfolio Approach classification was obtained, by indicating the number of companies which fall into a particular group. First of all, a distinction was made among firms which on the whole address poverty and those that do not. Among the firms which engage in poverty alleviation, it was investigated which concrete business approaches – including their variations – on poverty alleviation were pursued. Afterwards it was analyzed whether the employed initiatives were undertaken in isolation or in simultaneous combination. Lastly, based on the entire prior analysis the firms' approaches to poverty alleviation are classified. The resulting Poverty Alleviation Portfolio Approach builds upon the CSR framework presented in the theoretical part and thus firms can be classified as inactive, reactive, active or proactive. The following data collection, codification and classification parts outline in more depth the derivation of a firm's final classification.

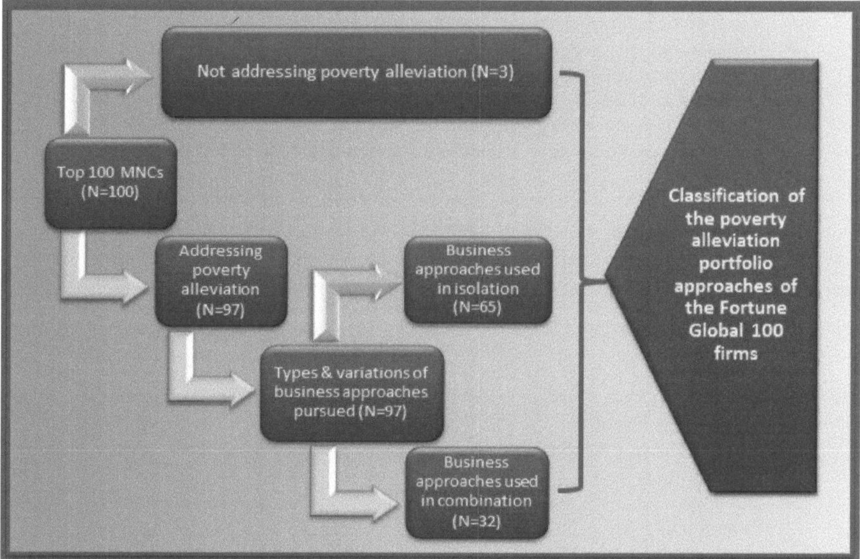

Figure 3-2: The Conceptual Research Model

3.1.6 Data Collection

Concerning the *data collection*, company quotes were collected from the *websites and CSR reports of the Fortune Global 100 firms*. The in-depth approach of the data collection is displayed in the Data Collection Methodology Section of Appendix F: Table 5-23 lists the variables to be collected whereas Table 5-24 displays a screen shot of the resulting Excel data collection sheet. To be more concrete, the data collection focused on receiving corporate information with regard to the *following variables*:

- Company information:
  - Company name
  - Country and region of origin

    - Company sector
    - Issue prioritization of poverty
    - Issue framing of poverty
- Overall statement of firm with regard to poverty alleviation
- Type and variation of undertaken business strategies on poverty alleviation:
    - Business-as-Usual
    - Classical and Strategic Corporate Philanthropy
    - Narrow and Broad Microfinance
    - Narrow and Inclusive Bottom of the Pyramid
    - Narrow and Broad Social Entrepreneurship
- Undertaken Cross-Sector Partnership approach on poverty alleviation:
    - None, limited, moderate or extensive partnering approach
- Information on whether the business strategies on poverty alleviation are combined:
    - Number of business strategies combined
    - Type of business strategies combined
- Information on whether these business strategies on poverty alleviation are undertaken in isolation or in partnerships:
    - Number of business strategies combined with Cross-Sector Partnerships
    - Type of business strategies combined with Cross-Sector Partnerships
- Additional useful poverty-related information and general comments

In order to obtain the relevant *company statements and quotes*, all websites – with a particular focus on CSR-related sections – were skimmed. Moreover, the website search function as well as the Google search function on particular websites was employed for 'poverty'. Furthermore, the entire reports were skimmed as well and the search function was applied within the report on the following *keywords*:

- Poverty
- Poor
- Business-as-Usual
- Foreign Direct Investment
- Philanthropy
- Donation
- Foundation
- Volunteering
- Microcredit
- Microfinance
- BOP
- Base of the Pyramid
- Bottom of the Pyramid
- Social Business
- Social Entrepreneur(ship)
- Social Enterprise
- Partnering
- Partnership

All relevant information obtained was consequently copied in Excel by indicating the source and – if originating from a report – the page number of the retrieved quote.

3.1.7 Data Codification

As a next step – to analyze the obtained data – the retrieved quotes were further *tabulated* in Excel to create an overview of the performance of the firm. Cross-tabulation among sub-groups for distinct firm characteristics, such as country and region of origin, sector and company narrative, was also provided due to their functions as possible explanatory variables, as explained later on. A detailed overview on how the data has been tabulated can be found in Table 5-25 until 5-35 within the Coding Methodology section of Appendix F. There, the *codification* options per variable are presented and the underlying reasoning per codification option is displayed. These guidelines should facilitate and objectify the codification process. However, it has to be noted that the actual company initiatives were not always as unambiguous as presumed in the tables and consequently in unclear or vaguely described cases the most applicable codification was employed. A frequently remarked criticism of coding is the fact that it results in the loss of detail, richness and context of information (Bryman and Bell, 2007). In order to avoid this and to illustrate the findings, *exemplary quotes* on the distinct categories are provided throughout the results section.

3.1.8 Data Classification

In order to further classify the firms based on the obtained information, the *classification framework* on the Poverty Alleviation Portfolio Approaches from the literature review was used. Firms were thus classified according to four categories, namely as inactive, reactive, active or proactive. In order to *quantify the data* for Excel calculations, an inactive sub-classification score was inserted with a score of 0, a reactive one with 1, an active one with 2 and a proactive classification with 3. More concretely, the ultimate Poverty Alleviation Portfolio Approach classification of the firms was determined based on the following seven different *sub-classifications scores*:

- The Corporate Philanthropy Classification;
- The Microfinance Classification;
- The Bottom of the Pyramid Classification;
- The Social Entrepreneurship Classification;
- The Cross-Sector Partnership Approach Classification;
- The Classification on the Combination of Business Strategies; and
- The Classification on the Combination of Business Strategies with Cross-Sector Partnerships

Based on these seven sub-classifications, a firm's overall *Poverty Alleviation Portfolio classification score* was calculated by summing up the seven sub-classification scores. A visualization of these components which jointly determines the overall classification can be found in Figure 3-3.

Figure 3-3: Poverty Alleviation Portfolio Classification Components

Due to the fact the maximum Poverty Alleviation Portfolio Approach score which could be obtained was 19 since 4 classification possibilities needed to be distinguished, the overall classification ranges for the added total scores were set with equal ranges as follows:

- *Score between 0 and 4:* ***Inactive*** *(No Poverty Alleviation Portfolio Approach)*
- *Score between 5 and 9:* ***Reactive*** *(Limited Poverty Alleviation Portfolio Approach)*
- *Score between 10 and 14:* ***Active*** *(Moderate Poverty Alleviation Portfolio Approach)*
- *Score between 15 and 19:* ***Proactive*** *(Extensive Poverty Alleviation Portfolio Approach)*

Once again, a detailed overview on the exact classification steps and possibilities can be obtained in the Company Classification Methodology section of Appendix F. Hereby, the Table 5-36 presents in greater depth the distinct classification components and options. Moreover, Table 5-37 provides an overview of an exemplary sub-classification process, while Table 5-38 displays the ultimate classification process.

3.1.9 Data Analysis Approach

In the proceeding results and discussion section, as a first step the *findings* on the corporate engagement in distinct poverty alleviation approaches as well as their variations and combinations are *described* in an observational manner. Afterwards an overall *discussion* of the main results is provided. In order to further dig into the research topic, subsequently a first *exploratory analysis* of the causal explanation of the differences in classification scores among companies is provided. This analysis serves solely as a first indication for potential influencing variables on a company's Poverty Alleviation Portfolio Approach. The suggested variables consequently need to be statistically tested to assume validation in further research.

3.1.10 Exploratory Analysis

Based on the data collected as well as on a subsequent literature search, three variables were selected as possibly explaining a firm's Poverty Alleviation Portfolio Approach, namely company sector, country/region of origin as well as a company's narrative on the issue of poverty. A visualization of the proposed relationship between the variables is displayed in Figure 3-4.

Company Sector and Region of Origin

To begin, *company sector* has been mentioned by numerous studies as directly impacting a firm's CSR reporting as well as the actual CSR initiatives undertaken (Wanderly et al., 2008; Roberts, 1992; Williams and Aguilera, 2008; Chochran and Wood, 1984). Moreover, the same assumption of a causal relationship between a firm's CSR reporting and engagement has been made with respect to a *company's country or region of origin* (Williams and Aguilera, 2008; Wanderly et al., 2008; Chapple and Monn, 2005).

Company Narratives on Poverty

Furthermore, another potentially relevant explanatory variable in this context would be a *company's overall narrative on the issue of poverty* (O'Dwyer, 2002). In general, a company narrative on CSR relates to a firm's perception on a certain matter and thereby also frequently outlines an organization's underlying reasoning for engagement in CSR activities (O'Dwyer, 2002). Precisely defining and measuring a firm's narrative on poverty would be beyond the scope of this thesis. Thus, based on the accessible data, two possible aspects, which might at least partially display a firm's narrative on poverty, were selected for further analysis, namely a firm's issue framing and a firm's issue prioritization approach with regard to poverty.

Issue Framing and Issue Prioritization

An *issue*, such as poverty, is defined as an "... unresolved subject of societal discontent that exists due to regulatory gaps; which involve great expectational gaps; leading to controversies; which (could) have an impact on the company and its reputation." (van Tulder and van der Zwart, 2006, p. 157). Subsequently, *issue framing* corresponds to the judgment and interpretation of an issue whereas *issue prioritizing* refers to the themes focused on an organization's CSR (or overall) activities (Mittal and Ross, 1998; Dawkins and Lewis, 2003). Literature has frequently highlighted the impact which issue prioritization as well as issue framing on CSR have on a firm's CSR reporting and CSR engagement (Dawkins and Lewis, 2003; Mittal and Ross, 1998; Wanderly et al., 2008). In addition, these findings have also stressed that a company's approach to CSR issue framing and prioritization is highly impacted by the respective company sector or origin (Dawkins and Lewis, 2003; Williams and Aguilera, 2008). Thus, it can be expected that company size and country or region of origin do impact the firm narrative on poverty, which in turn would influence a company's Poverty Alleviation Portfolio Approach.

Proposed Explanatory Model

As a result of the prior discussions, as displayed in the Figure 3-4, company sector and country or region of origin is proposed as *independent variables* which directly influence the *dependent variable* 'Poverty Alleviation Portfolio Approach'. In contrast, company narrative on poverty, measured via issue framing and issue prioritization, can be considered a possible *mediating variable* between the independent variables of company sector as well as region or country of origin and the dependent variable 'Poverty Alleviation Portfolio Approach'. A well-known definition of a mediating variable is provided by Baron and Key (1986, p. 1176): "In general, a given variable may be said to function as a mediator to the extent that it accounts for the relation between the predictor and the criterion. [...] Whereas moderator variables specify when certain effects will hold, mediators speak to how or why such events occur." This reasoning implies that a mediator variable has the potential to explain the relationship between the independent and the dependent variables, which appears to be applicable with regard to the company narrative on poverty.

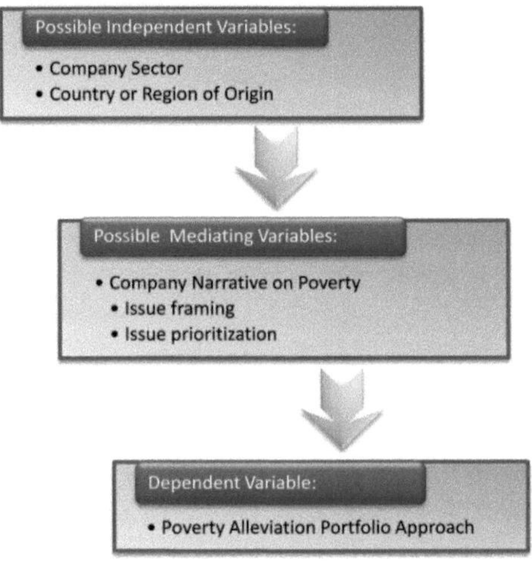

Figure 3-4: Proposed Explanatory Model for a Companies' Poverty Alleviation Portfolio Approach

Methodology for the Exploratory Variables

Concerning the employed methodology, information on *country or region of origin* and *company sector* has already been collected, as outlined previously, and no further codification of information was required. However, company narrative was a novel variable of analysis, which still needs to be codified. In this context, a company's narrative was comprised of two aspects, namely a firm's issue prioritization and its issue framing approach on poverty.

Information on *issue framing* was retrieved from the provided company statements on poverty alleviation. A common way of framing issues among academics is to look at the tone in which an issue is presented (Mittal and Ross, 1998). On the one hand, an issue can be framed in a negative tone, by regarding the issue as a threat to a firm's operation, while highlighting the necessity or obligation to combat this issue to hinder further harm. On the other hand, an issue could be expressed in a positive tone, by stressing its opportunities for companies and by focusing on its solutions or possible contributions to addressing it (Mittal and Ross, 1998). Furthermore, an issue could be presented in neutral terms which would imply that either no explicit positive or negative issue-framing tone from the statement can be retrieved or that the statement contains both positive and negative framing aspects. Lastly, an issue might not be framed at all. Accordingly, being based on the poverty statements collected before, the following issue framing approaches were distinguished:

- Issue of poverty is not framed at all
- Issue of poverty is framed in negative terms
- Issue of poverty is framed in neutral terms
- Issue of poverty is framed in positive terms

In turn, data on *issue prioritization* has been derived from the gathered information on companies' CSR focus themes and issues. All these themes were reviewed to identify which firms emphasize poverty as their CSR priority. Hereby, poverty was once again defined in broad terms, which implies that related phrases such as 'community development' or 'social inclusion' would also fall under a poverty focus. As a result, two issue prioritization approaches could be differentiated:

- Issue of poverty is prioritized
- Issue of poverty is not prioritized

3.1.11 Research Characteristics

Key characteristics of advanced qualitative research are its reliability, replication, validity and confirmability and hence these dimensions must be considered in the research design (Bryman and Bell, 2007).

Research Reliability

To begin, *reliability* refers to "... the question of whether the results of a study are repeatable" (Bryman and Bell, 2007, p. 40). Repeatability of this research can be partially presumed due to the fact that companies are likely to continue web-based reporting of their CSR-related activities, especially as the importance of CSR as well as the Internet usage rates continue to increase. Nonetheless, while past reports usually remain accessible, alteration of websites can be expected as contents are generally updated on a regular basis. Hence, while the research approach itself can be repeated, the results obtained at a future period might be slightly different. Reliability is also related to representativeness which – as outline above – is likely to be high for this study due to the relatively large sample size of 100 firms, which in turn should increase the reliability of the research.

Research Replication

Following, *replication* encompasses the possibility of a future replication of a study by another researcher and hence is closely related to the research reliability (Bryman and Bell, 2007). Replication of this research is possible since the concrete steps undertaken are outlined in depth in the methodology section, which should in turn enable future replication.

Research Validity

Next, *validity* relates to "... the integrity of the conclusions that are generated from a piece of research" (Bryman and Bell, 2007, p. 40). This research can be considered valid because content analysis of corporate reporting has been used in past studies as well, to derive information on the initiatives of firms with regard to CSR matters. Besides, internal validity is presumed due to the strong linkage of the theoretical and practical parts of the paper through the classification model. Moreover, external validity is obtained through the large and diverse sample size which should enable a "... generalization across social settings" (Bryman and Bell, 2007, p. 410).

Research Confirmability
Lastly, *confirmability* refers to the objectivity of a study. In this regard, Bryman and Bell (2007, p. 414) remark: "While recognizing that complete objectivity is impossible in business research, the researcher can be shown to have acted in good faith." In the context of this study it should be emphasized that codifications and classifications of obtained data are in general partially subjective which is also recognized as a main limitation of this research approach. In order to ensure confirmability, substantial attention has been paid to explicitly specifying the codification and classification steps and reasoning in the prior methodology section. It is nonetheless conceded that a slight subjectivity to the results could not be avoided.

3.2 Results

This section displays the results of the primary research on Fortune's largest 100 MNCs. Hereby, the actual initiatives of the firms in term of poverty alleviation will be presented, in particular by outlining which approaches have been employed and whether these approaches were undertaken in isolation or in combination. Based on these findings, the companies will be classified in terms of their Poverty Alleviation Portfolio Approaches.

3.2.1 Overall Observations

To begin, some general observations from the data collection need to be stated.

Concerning the data gathering, English-language websites could be accessed for all 100 firms in the sample. Moreover, almost all companies provided a CSR-related section on their websites. In contrast, obtaining CSR reports was more difficult due to the fact that the reports were kept in various places on the websites and possessed varying titles, such as CSR Report, Sustainability Report or Corporate Citizenship Report. For simplicity matters, all these distinct report types will be jointly referred to as 'CSR reports' in the proceeding sections. Overall, out of the 100 firms, 88 had a specific CSR report. And out of these 88 firms, 4 companies had more than one CSR report. For the remaining 12 firms, 4 had a joint annual and CSR report while 8 companies did not provide any CSR related report on their websites. Hence, reports from 92 companies could be reviewed. The goal was to obtain the most recent CSR Report per firm but since not all firms published their latest versions, the investigated reports originated from 2009, 2010 or 2011. To be concrete, 12 reports were published in 2011, 73 reports in 2010 and 7 reports in 2009. The reviewed company websites and reports can be found in the bibliography list.

Generally speaking, with regard to online reporting by the examined firms, the obtained information was plentiful. However, it turned out to be challenging to particularly obtain poverty-related information – be it on the websites or within the CSR reports. The underlying reason is that companies tend to organize their CSR activities foremost either according to the stakeholder groups addressed, such as environment, customer or employee initiatives, or according to the fields of engagement, for instance, corporate governance, human rights or safety. Poverty-related undertakings can thus be found under distinct stakeholder categories or engagement fields. As an example, as stakeholders 'the poor' can be classified as community members, employees, customers or suppliers. Focus on the issue addressed, such as poverty, was much less common. Hence, the data gathering had to be conducted very extensively in order to ensure that no relevant information was overlooked. As a general tendency, poverty was commonly discussed in particular related to community initiatives.

On a methodological note, it has to be remarked that 'poverty' was defined in a broad manner with regard to the data collection. Of course, foremost attention was paid to statements which directly contain the word 'poverty'. However, also statements which relate to poverty indirectly or use a highly related term were also incorporated into the analysis. Examples of such related terms include 'social exclusion', 'the vulnerable', 'the marginalized' or 'the local communities in developing countries'.

It is also highly important to stress that this research aimed at classifying companies solely with regard to their *poverty* approaches not on their entire *CSR* portfolios. Hence, one cannot presume a correlation between poverty initiatives and CSR initiatives in general per firm. For instance, Honda, which clearly classified as inactive with regard to poverty, is highly engaged in other CSR activities, being related to the environment and road safety (Honda website, 2011). Thus, the conducted research has generally demonstrated that poverty was oftentimes not stated as focus area of a firm's sustainability initiatives. The CSR focus themes vary from firm to firm but some examples of other frequently addressed matters include environmental degradation, corporate governance, health and safety or human rights and diversity.

In the proceeding parts of this section, the main findings per examined variable are outlined and illustrated with overview diagrams and exemplary quotes. The number of quotes has been limited to approximately three per codification in order to present a brief illustration on the variety of actual company initiatives.

3.2.2 Provision of a Poverty Statement

First of all, the research results reveal that out of the 100 reviewed companies, 49 percent do provide a poverty or *poverty-related statement* on their website or CSR report. A visualization of the concrete results in this regard and some exemplary quotes can be found in Figure 3-5 and Table 3-1. In general, it was mostly stated that addressing social problems, such as poverty, is of importance for companies. This supports the assumptions made in the theoretical part, that MNCs are regarded as important actors for addressing poverty (see ING Group's quote below). However, firms also tend to examine these issues from a business perspective and thus also tend to emphasize that combating poverty is not solely required from a societal perspective but also from a business perspective in order to successfully operate as firms in the future (see for instance Wal-Markt's and Gazprom's quotes below).

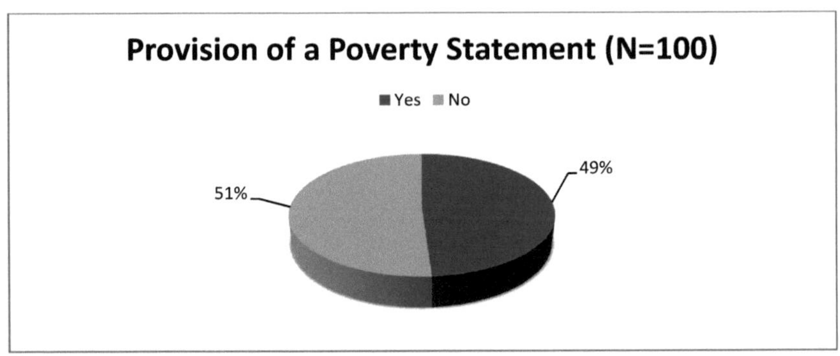

Figure 3-5: Company Overview Diagram on the Provision of a Poverty-Related Statement

Company	Exemplary Quotes
Wal-Mart Stores	"Some of the biggest challenges facing the world right now are around *poverty* and hunger, global population growth, resource management and conservation. Any company that wants to lead in the future will have to address these issues." (Company Website, 2011).
ING Group	"Being a good corporate citizen means that we want to contribute to positive change in society. We are focused on finding innovative solutions that address the local and global challenges our customers face, like climate change, *poverty* or ageing of the population. [...] We want to ensure that we remain responsive to global issues such as biodiversity, global warming, the aging population and poverty." (Company Website, 2011).
Gazprom	"It is essential to eliminate *poverty* and existing wealth inequalities if we want to ensure our planet's peaceful development." (Company Website, 2011).

Table 3-1: Exemplary Company Quotes on the Provision of a Poverty-Related Poverty Statement

3.2.3 Engagement in Business-as-Usual

As discussed in the theoretical part above, engagement in *Business-as-Usual* with regard to poverty would imply that a firm does not engage in any poverty-related activities. However, prior research has highlighted the fact that companies are unlikely to report on initiatives they *do not* undertake but instead tend to focus on the activities they *do* pursue (Belal and Cooper, 2007). Hence, it is unlikely to find a quote, which directly stated that a firm does not engage in any poverty-related approaches. Indeed, no such quote could be obtained, with the statement from Berkshire Hathaway below being the most related one, by emphasizing a focus on economic – and not social – goals. In contrast, only quotes could be obtained which highlighted the general impact of business operations on development without taking a clear position on it or quotes which emphasized that engagement in Business-as-Usual is not sufficient (see Table 3-2). As a result, this category was analyzed slightly differently. Responding to reasoning above, it can be partially presumed that if a firm does not report on a manner it does most likely not undertake it or prioritize it. To put it into the words of Belal and Cooper (2007, p. 18): "There is an absence of reporting of bad news. From this we could surmise that where there is a lack of good news there will be a subsequent lack of CSR reporting." Hence, the firms on which no information on any poverty-related initiatives could be obtained were classified as pursuing a Business-as-Usual approach. Accordingly, as visualized in Figure 3-6, three firms classified as complete inactive with regard to the issue on poverty, namely Berkshire Hathaway, American International Group and Honda Motor. However, as mentioned above, Honda Motor was active with regard to non-poverty related CSR initiatives while for Berkshire Hathaway and American International no CSR Report or any other CSR-related information could be obtained on the companies' websites. Due to the fact that these three firms already possess a classification score of 0 on all other variables for the overall Poverty Alleviation Portfolio Approach score, it has been decided to exclude the Business-as-Usual variable from the overall classification calculation.

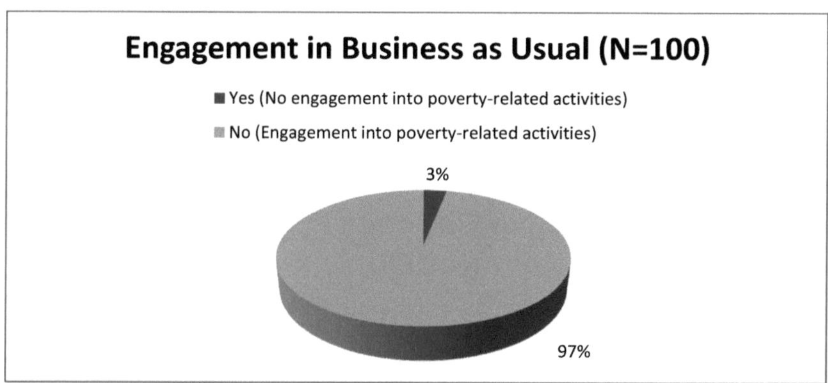

Figure 3-6: Company Overview Diagram on the Engagement in Business-as-Usual

Company	Exemplary Quotes
Statement on Engagement beyond Business-as-Usual	
BP	"We understand that *business-as-usual* is not an option, and we are making substantial changes to the way we work." (Company Website, 2011).
Statement on Importance of Usual Business Activities for Poverty Alleviation	
Glencore International	"Glencore's global presence and economic strength has a positive impact on the communities in which we operate. We provide *employment* and numerous other benefits, thereby contributing directly and indirectly to the prosperity and *development* of our host countries in general and local communities in particular. [...] When examining the *economic impact* of our industrial activities in our host countries, one of the most obvious effects is the employment we create." (Company Website, 2011).
BASF	"*Economic growth* is a precondition to achieve the Millennium Development Goals. Therefore, the biggest contribution of companies like BASF is economical commitment in developing and emerging countries, mainly *Foreign Direct Investments.*" (Company Website, 2011).
(Approximate) Statement on Engagement in Business-as-Usual	
Berkshire Hathaway	"We give each [manager] a simple mission: Just run your business as if: 1) you own 100% of it; 2) it is the only asset in the world that you and your family have or will ever have; and 3) you can't sell or merge it for at least a century [...] Our long-term economic goal is to maximize Berkshire's *average annual rate of gain* in intrinsic business value on a per-share basis." (Company Website, 2011).

Table 3-2: Exemplary Company Quotes on Business-as-Usual

3.2.4 Engagement in Corporate Philanthropy

With regard to *Corporate Philanthropy* (CP) initiatives on poverty, the sample firms turn out to be extremely active with 93 out of a 100 engaging in CP activities as displayed in Figure 3-7. Among these 93 firms, 42 per cent engage in classical CP while 58 per cent pursue strategic CP (see Figure 3-8). As revealed by the company citations in Table 3-3, the concrete forms of engagement, even within these variations, are diverse. Most companies provide financial corporate donations while also engaging in employee giving and volunteering. Beyond that, firms which engage in strategic CP also donate non-financial goods, provide training and expertise and establish their own foundations which also focus on poverty-related initiatives. The number and scope of undertaken initiatives is more extensive for the strategic approach in comparison to the classical variation. Moreover, the former has a more strategic orientation than the latter by being more closely related to a firm's core business and area of expertise. Ultimately, engagement in partnerships is more common under the strategic CP approach as well.

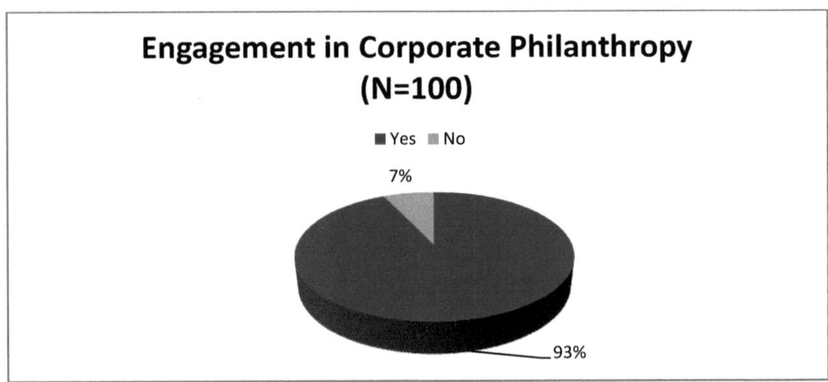

Figure 3-7: Company Overview Diagram on the Engagement in Corporate Philanthropy

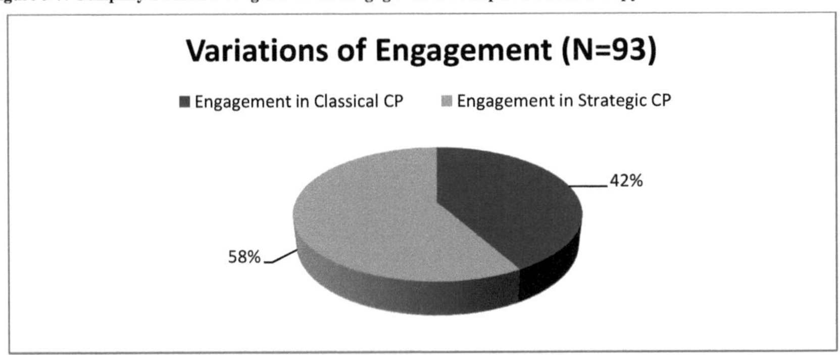

Figure 3-8: Company Overview Diagram on the Variations of Engagement in Corporate Philanthropy

Company	Exemplary Quotes
Statement on Engagement in Classical Corporate Philanthropy	
Volkswagen	"Workforce *donations*: The "Starthilfe" (Getting Started) initiative is a project devoted to combating the growing problem of child poverty in the Wolfsburg region. "Starthilfe" uses donations to launch, promote and focus projects and measures to alleviate child *poverty*. Also in 2010, the workforce donated almost €1 million to help the victims of the natural disasters in Pakistan and Haiti." (Company Website, 2011).
AT&T	"AT&T invests significant *resources* to advance education, strengthen communities and improve lives. Through *philanthropic initiatives*, AT&T and the AT&T Foundation support projects that create opportunities, make connections and address *community needs* where we — and our customers — live and work." (Company Website, 2011).
Lloyd Banking Group	"Lloyds Banking Group is the biggest corporate investor in UK *communities*. Last year, we *invested* £76 million in communities, including support for *financial inclusion*. [...] In addition, we made a very significant investment in the provision of social bank accounts, to help bring those who are excluded from mainstream financial services into the financial system." (Company Report, 2010, p. 4).
Statement on Engagement in Strategic Corporate Philanthropy	
Carrefour	"The Carrefour Foundation aims to provide the *vulnerable* with access to safe and varied foodstuffs, particularly through its *product donations* and in-store collection campaigns." (Company Report, 2010, p. 51). "Evidence of Carrefour's social commitment is also seen in the work of its *Corporate Foundation* and the numerous local initiatives in the fields of food program, professional integration and international solidarity." (Company Report, 2010, p. 44). "For 10 years, the Carrefour Foundation has been supporting the work of *local non-profit associations* in aid of the *disadvantaged*. Its various actions focus on three missions that reflect the Group's business lines, the skills of its employees and its operational scope. [...] The Carrefour Foundation aims to help improve the employment prospects of those suffering from *social exclusion* by supporting microcredit, rural development, and integration programs. [...] With an annual budget of €4.5 million, the Carrefour Foundation has supported more than *220 projects in 40 countries* since it was first set up, including 52 in 2010. For example, the Foundation is the largest private donor to the Food Bank Federation." (Company Report, 2010, p. 51) "In addition to the projects implemented [...], there are some in which the companies' participation consists not of cash donations, but of providing material goods or direct services. Workers have been involved in a number of initiatives by collecting money or goods or doing *voluntary work* to help the *less fortunate*." (Company Report, 2010, p. 114).

Company	Exemplary Quotes
ENI	"As part of its commitment to creating sustainable value, Eni has established *Eni Foundation*, a foundation with its own goals, financial resources and operating tools, whose mission is to promote and carry out solidarity initiatives both in Italy and abroad, in support of the *disadvantaged and vulnerable*, especially children and the elderly. Eni Foundation operates primarily by developing its own projects, which are aimed at protecting the health and promoting the well-being of children, adolescents and the elderly, improving their quality of life and *mitigating hardship and social exclusion*. In order to implement its programs, the Foundation can set up cooperation initiatives and partnerships, both in the planning and implementation phases, with NGOs, humanitarian associations, as well as local institutions and government agencies. *Partners* are selected on the basis of specific and proven skills and competences in the areas of interest and according to the complexity of the projects being undertaken." (Company Website, 2011).
Lukoil	"Social and charity programs are inherent to the Company's corporate strategy and help promote *constructive cooperation* with the governmental authorities, business community and the public. Corporate programs are *target-oriented* and use the professional expertise and human potential of the regions. However the Company has a clear understanding that charity should not become fertile ground for social parasitism. Therefore, alongside with the conventional forms of charity, LUKOIL implements its *strategic charity* and social investment programs in such a way that there is always a link between social problems and the Company's strategic goals. The pre-requisite for such an approach is cooperation between the commercial, non-profit and public sectors aiming to solve urgent *social and economic problems of the local communities.*" (Company Website, 2011).

Table 3-3: Exemplary Company Quotes on Corporate Philanthropy

3.2.5 Engagement in Microfinance

Engagement in Microfinance is not as widespread among firms as engagement in Corporate Philanthropy with only 31 firms pursuing this strategy (see Figure 3-9). As revealed in Figure 3-10, out of these 31 firms 42 per cent engage in the narrow Microfinance variation whereas 58 per cent undertake the broad Microfinance approach. Similarly to the quotes presented in Table 3-4, the narrow Microfinance variation usually consists of the provision of Microfinance loans without additional training or capacity building initiatives. In contrast, with regard to broad Microfinance additional services are provided together with support and training activities. Overall, the number and extensiveness of the projects for the broad variety is higher with the initiatives being more closely related to the firm's core business. Furthermore, commitment to Microfinance partnerships is more likely. As a final note, as already presumed, the sector appears to be an influential variable for the Microfinance strategy, with financial firms especially likely to be actively engaged in Microfinance. To be concrete, out of the 27 financial firms, 11 firms follow a broad approach, 7 a narrow one and solely 9 financial firms do not engage in any Microfinance initiatives.

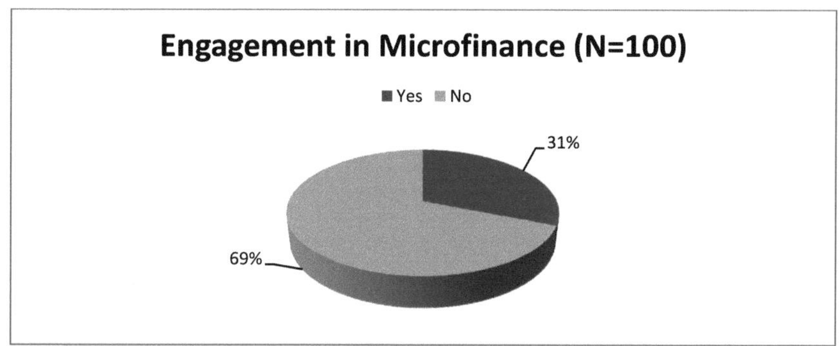

Figure 3-9: Company Overview Diagram on the Engagement in Microfinance

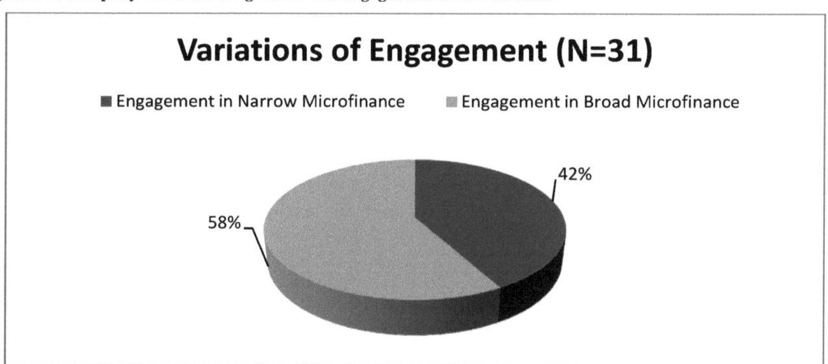

Figure 3-10: Company Overview Diagram on the Variations of Engagement in Microfinance

Company	Exemplary Quotes
	Statement on Engagement in Narrow Microfinance
Banco Santander	"Santander *Micro Credits* encourages growth in small businesses, thereby enabling the *least favorable sectors of society* to merge from *poverty*. Loans are mainly made to informal micro companies that are unable to obtain loans. More than 80% of micro credits go to businesswomen who, in groups of five to six, receive loans of an average amount of between EUR 300 and EUR 500 without the need for additional guarantees. In 2010, the volume of loans reached BRL 123 million. Since 2002, a total of BRL 780 million has been granted." (Company Report, 2010, p. 29).
China Mobile Communications	"In Fujian, in 2008, we started a pilot project for rural financial information service in Ningde city, and developed the Rural *Microcredit* Self-service Information Platform. We consolidated agriculture-related resources of financial institutions, *poverty alleviation* agencies, agricultural businesses and the government to achieve centralized management of data, information and microcredit service. [...] We are planning to expand this service throughout the province in 2011." (Company Report, 2010, p. 26).

Company	Exemplary Quotes
Repsol YPF	"The Repsol Foundation carries out activities in countries where the company has a presence, with specific projects designed to suit the requirements of each area. In Ecuador it is running a *microcredit* program in the provinces of Orellana and Sucumbíos for *women on low incomes*, enabling them to generate their own employment resources." (Company Website, 2010, p. 30).
	Statement on Engagement in Broad Microfinance
Chevron	"*Microfinancing* provides Indonesians with business opportunities. The programs we support, through the delivery of *low-cost loans*, reach a variety of enterprises outside of our operations. "Microfinancing is about enabling people to build businesses and employ others, resulting in stronger communities," said Ted Etchison, Chevron senior vice president for operations in Kalimantan. "Microfinance is not charity. It's about building *capability and empowerment*, and it places the responsibility for success on the participants. We help plant the seed. The people then develop their own livelihoods. In East Kalimantan and West Java, we partnered with government-owned financial institutions Permodalan Nasional Madani and Baitulmaal Muamalat to form the Community Enterprise Development program, offering access to low-cost loans and management training to community-based business groups and small businesses. [...] We helped establish farmers' networks, where farmers turn idle land into fields of abundant crops. We provide *training and support* and share knowledge with each other." (Company Report, 2010, pp. 34-35).
Crédit Agricole	"The Crédit Agricole Grameen Foundation was set up in September 2008 on the initiative of Crédit Agricole S.A., in *partnership* with Professor M. Yunus, the founder of Grameen Bank and Nobel Peace Prize winner. Its role is to support *microfinance* institutions providing financial services to the *poorest*. This process has proved its effectiveness, particularly through microcredit to finance revenue-generating activities. Crédit Agricole has provided the Foundation with €50 million and twelve specialists so that the Foundation can offer *loans, guarantees, equity and technical assistance* to microfinance institutions. It also encourages social business projects to provide access for the poorest to essential goods such as food, water, energy, education and healthcare. Based on both financial and social research, the Foundation provides funding to microfinance institutions that are intermediate in size and have demonstrated quality management and effective action to help the least well-off." (Company Report, 2010, p. 164). "Crédit Agricole has also formed several national *partnerships* with specialised non-profit organisations such as France Initiative, France Active, and Adie. Since October 2003, Crédit Agricole has partnered France Initiative, the leading financing network in France and Europe for entrepreneurs and for a local development system, generating thousands of jobs every year. France Initiative provides loans on trust to entrepreneurs with high potential projects to overcome their lack of capital and assist them throughout the start-up process. These start-ups generally have an above-average survival rate over three years of 83 per cent. By supporting microcredit organisations we help to give the *poorest* people access to financing, enabling them to grow their businesses and lift themselves out of *poverty*." (Company Report, 2010, p. 112).
Daimler	"Together with the non-profit *partner* CARE, Daimler Financial Services is supporting with an annual global Christmas donation integrated *microfinancing* programs to provide *poverty-fighting* projects. We offer *training* courses designed to show the people who receive such loans how to set up and successfully manage their own small businesses. At the same time, *assistance* is also provided to already existing microenterprises that are being hindered from making a profit by a lack of good business management skills." (Company Website, 2011).

Company	Exemplary Quotes
GDF Suez	"Microfinance Programs: 1) Entreprendre en banlieue (Business in the Suburbs): Founded in 2006, the goal of this program is to make unemployed adults from sensitive urban areas (ZUS) aware of entrepreneurship by creating ADAMs (*micro-entrepreneur* awareness and support associations) in these districts. This association network seeks out future micro-entrepreneurs, supports and assists them as they set up their businesses, and guides them as they look for funding by introducing them to local microcredit; 2) FinanCités: FinanCités is a venture-capital company with a mission to support micro-businesses in urban districts by providing them with additional financing at a crucial stage in their development. By providing its *financial help* to growing micro-businesses, FinanCités aims to contribute to the growth and permanent survival of small businesses in urban districts, help create jobs, provide a boost to the local economy and, more generally, *improve living conditions* in these urban districts; 3) MicroWorld, solidarity-based microcredit on the internet: MicroWorld aims to develop financing for micro-entrepreneurs throughout the world in the form of loans through the internet. Microcredit and microfinance efforts have shown that small, well-targeted resources can be used to initiate *sustainable change* that benefits everyone." (Company Website, 2011).

Table 3-4: Exemplary Company Quotes on Microfinance

3.2.6 Engagement in Bottom of the Pyramid

Out of the 100 companies examined, only 13 engage in *Bottom of the Pyramid* (BOP) initiatives, as visually displayed in Figure 3-11. In turn, out of these 13 firms, 62 per cent follow a narrow approach and 38 per cent pursue an inclusive BOP strategy (see Figure 3-12). Overall, the distinction between the two variations is relatively clear cut. While the narrow approach regards the poor primarily as customers to whom goods are sold, the inclusive strategy aims at involving the poor at additional stages of a firm's value chain – be it as suppliers, entrepreneurs or employees. Moreover, once again under the inclusive BOP strategy, the undertaken activities are more elaborate and plentiful in number and are placed at the core of a firm's strategic orientation. Engagement in BOP partnerships also appears to be more likely. The exemplary quotes in Table 3-5 should thus serve to further illustrate the distinct strategies.

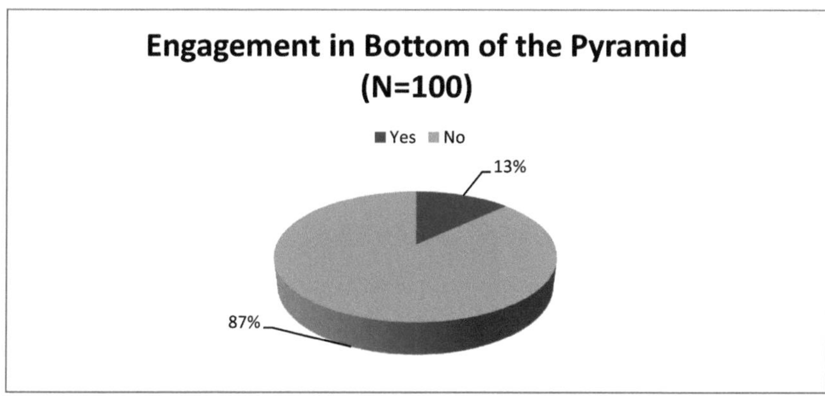

Figure 3-11: Company Overview Diagram on the Engagement in Bottom of the Pyramid

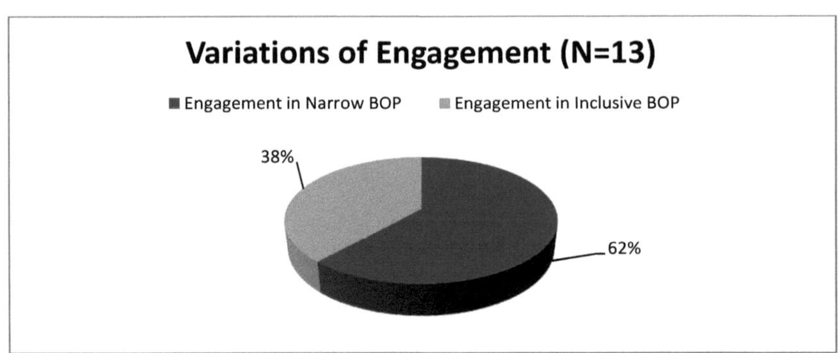

Figure 3-12: Company Overview Diagram on the Variations of Engagement in Bottom of the Pyramid

Company	Exemplary Quotes
	Statement on Engagement in Narrow Bottom of the Pyramid
Telefónica	"Regarding the *base of the pyramid segment*, Telefónica continued offering *products* that were adapted to their economic needs. In Brazil, Telefónica offers 'People's Broadband', aimed at a *low-income population* with no monthly fee and no telephone line rental, with more than 200,000 customers. And in Peru, Telefónica offers the "Fono Ya" fixed line service with wireless access which has around 470,000 *customers in the lower socioeconomic segments.*" (Company Report, 2010, p. 83).
Panasonic	"BOP *energy products for poor*: SANYO aims to spread beneficial products that are *affordable* for people in un-electrified areas and support the *improvement of their lifestyles*. The solar LED lantern is our first product to help realize "electricity" to regions without electricity, and "light" to regions without light." (Company Website, 2011).
Sony	"Sony is particularly aware that emerging economies face significant *development challenges* and is exploring new business approaches to address them. Efforts to date include inviting experts from outside the Company to hold seminars for pertinent employees with regard to promoting *BOP business.*" (Company Website, 2011).
	Statement on Engagement in Broad Bottom of the Pyramid
Repsol YPF	"*Inclusive businesses* are economic activities that enable the *most disadvantaged groups* in society to become involved in generating value. In short, these are financially profitable business initiatives that use a *mutually beneficial* structure to incorporate *low-income communities* into their *value chains* and improve their *quality of life*. Over the course of 2010, the purchasing and contracts unit worked with the Repsol Ecuador Foundation to develop this model, enabling us to identify *inclusive suppliers*, with whom we contract to supply various products and services for the company." (Company Report, 2010, p. 37).
J.P. Morgan	"J.P. Morgan's Social Finance (SF) unit provides investment and capital markets services to *social enterprises*, funds, foundations, non-governmental organizations (NGOs), development financial institutions and other investors serving the *base of the economic pyramid*. In doing so, SF achieves a *double bottom line* of social benefit and financial return. [...] The *'Base of the pyramid'* describes groups of people in emerging markets who earn less than $3,000 a year." (Company Website, 2011).

Company	Exemplary Quotes
BP	"We run programs to develop *local supply chains* and build *business skills* locally to help companies meet the standards needed to supply us and other clients. At the same time, we benefit from the local sourcing of goods and services. A number of our sites are working towards increasing the amount of goods and services sourced from local or national *suppliers*. A number of our major operating sites are working to improve representation from their host country in their workforce. At 87% of major operating sites surveyed in 2010, the national employees are now represented on the senior management team." (Company Website, 2011).

Table 3-5: Exemplary Company Quotes on Bottom of the Pyramid

3.2.7 Engagement in Social Entrepreneurship

Comparable to the Bottom of the Pyramid initiatives, engagement in *Social Entrepreneurship* (SE) is also very limited, with 13 out of 100 companies pursuing this approach (see Figure 3-13). Figure 3-14 further reveals that out of these 13 firms, 46 per cent undertake a narrow and 54 per cent a broad SE variation. Classifying SE approaches is rather challenging due to the diversity of undertaken initiatives, as indicated in the exemplary quotes in Table 3-6. Engagement in SE has thus been classified as broad if it was undertaken by the company itself, with the activities recognized as plentiful and highly related to poverty. In turn, it was considered as narrow when only support to external parties had been provided and the initiatives were limited and only loosely related to poverty. Furthermore, a broad SE approach once again implies a closer tie to a firm's core business and an increased likelihood of a Social Entrepreneurship partnering approach.

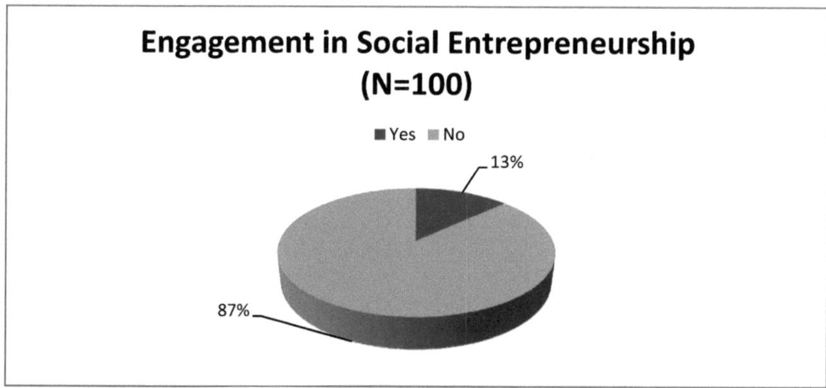

Figure 3-13: Company Overview Diagram on the Engagement in Social Entrepreneurship

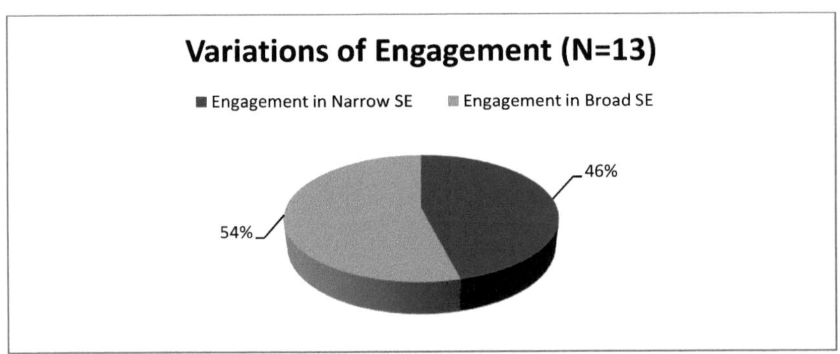

Figure 3-14: Company Overview Diagram on the Variations of Engagement in Social Entrepreneurship

Company	Exemplary Quotes
	Statement on Engagement in Narrow Social Entrepreneurship
BNP Paribas	"The Group has taken many other practical initiatives to promote *equal opportunity, social entrepreneurship*, discovery and learning. In 2010, a total of EUR 15 million was devoted to actions taken under the Projet Banlieues umbrella, including EUR 4 million in the form of *grants* and EUR 11 million in equity and debt financing." (Company Report, 2010, p. 17).
GDF Suez	"The Group has also created an international *social program* entitled 'Rassembleurs d'Énergie' [Energy Gatherers]. Its aim is to support social entrepreneurs working to provide access to energy and essential services for *poorer populations*. The program includes the creation of a *social fund* to finance innovative projects on a social and environmental level as well as a substantial amount of skills sponsorship for employees. Through this initiative, GDF SUEZ confirms its strong commitment to social action." (Company Report, 2010, p. 3).
Crédit Agricole	"It also encourages *social business projects* to provide access for the *poorest* to essential goods such as food, water, energy, education and healthcare [...] The Foundation also offers partial guarantees to enable these institutions to raise *finance* from Local Banks." (Company Report, 2010, p. 164).
	Statement on Engagement in Broad Social Entrepreneurship
BASF	"BASF SE and Grameen Healthcare Trust establish the Joint Venture BASF Grameen Ltd. in Bangladesh. The purpose of the company is to improve the health and business opportunities of the poor of Bangladesh. BASF is the first DAX30 company and the first chemical company in the world to set up a *social business with Grameen.*" [...] "BASF Grameen Ltd. combines business sense with social needs" Dr. Muhammad Yunus, Nobel Peace Prize Laureate and Managing Director of Grameen Bank [...] "Social business is an excellent way of creating value from values. Our market-oriented joint venture will provide long-term help in addressing *social challenges* in Bangladesh." Jürgen Hambrecht, Former CEO BASF SE. [...] "The social business serves a social purpose, covers its own costs and recoups the partners' initial investment. Any additional profits are *reinvested* fully in the company." (Company Website, 2011).

Company	Exemplary Quotes
Hewlett-Packard	"The Schwab Foundation for *Social Entrepreneurship* is a dynamic community of social entrepreneurs, driving measurable social improvement around the globe. In 2010, HP began partnering with Schwab Foundation to connect social entrepreneurs with HP researchers and information *technology experts*, and to share the methodology, processes, and tools HP uses to innovate *and problem solve.*" (Company Report, 2010, p. 122).
Royal Bank of Scotland	"Through *partnerships* with leading organizations like The Prince's Trust, we focused on supporting enterprise and entrepreneurship, and we backed this up with financial support to a range of new businesses and social enterprises." (Company Report, 2010, p. 5). "RBS Romania continued its partnership with NESsT in 2010, sponsoring its second *Social Enterprise Competition*. NESsT worksto solve social problems in emerging market countries by developing and supporting social enterprises to strengthen both their financial sustainability and their social impact. [...] Last year, RBS Community Banking launched a new partnership with Key Fund Yorkshire, a *community development* finance institution (CDFI) that has been providing investment to social enterprises and *community groups* for the last decade. During this time, it has invested more than £21 million, helping 125 business start-ups, creating 736 jobs and safe-guarding a further 574 jobs." (Company Report, 201o, p. 26).

Table 3-6: Exemplary Company Quotes on Social Entrepreneurship

3.2.8 Engagement in Cross-Sector Partnerships

In turn, engagement in Cross-Sector Partnerships was widespread with 73 out of the 100 firms pursuing it (see Figure 3-15). Hereby, as demonstrated in Figures 3-16 and Table 3-7, three variations of engagement can be distinguished, and are recognized as limited, moderate or extensive partnering approaches. Accordingly, out of these 73 companies, 36 per cent pursued a limited partnering approach, 42 per cent a moderate partnering approach and 22 per cent an active partnering approach. Distinctions among approaches were once again challenging due to the great diversity of initiatives undertaken. Overall, an extensive partnering approach is characterized by four or more projects with three or more partners, duration of more than two years in at least three countries and with financial and non-financial firm contributions. In contrast, a moderate partnering approach implies engagement in two or three projects with approximately two partners and a time span of one to two years. Activities should take place in two to three locations with the form of engagement being primarily financial. Lastly, a limited partnership approach encompasses one project in one country with one partner, duration of less than a year and the firm provides only financial contributions. Unfortunately, in reality the distinctions among approaches were not always as clear-cut as presumed and hence in the case of ambiguous cases the most applicable variation was selected.

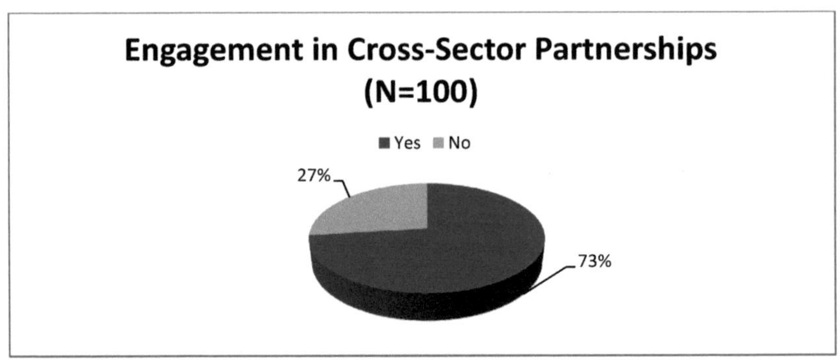

Figure 3-15: Company Overview Diagram on the Engagement in Cross-Sector Partnerships

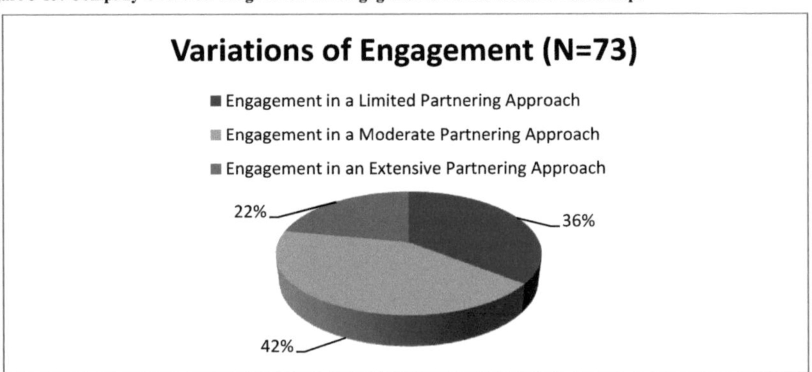

Figure 3-16: Company Overview Diagram on the Variations of Engagement in Cross-Sector Partnerships

Company	Exemplary Quotes
Statement on Engagement in a Limited Cross-Sector Partnership Approach	
General Electrics	"ILBF is engaged in a project — with support from the GE Foundation—to develop a network of companies that can work collectively to enable *poverty reduction* through core business strategies, and that can address some of the constraints to business in *emerging markets*. Working in *partnership* with locally established international partners, IBLF is piloting a program to support the development of a responsible business network in Tanzania." (Company Website, 2011).
Ford Motor	"Ford has been supporting *community efforts* since our founding more than 100 years ago. For us, it is not just about donating money. It is also about building *partnerships* and working with others to address the difficult challenges so many people are facing. This includes helping feed hungry people, providing mentors in classrooms and teaching teenagers to drive more safely." (Company Website, 2011).
Industrial and Commercial bank of China	"The Bank *cooperated* with China Foundation of Poverty Alleviation and donated RMB 900,000 to establish the first "ICBC – New Great Wall Senior Secondary Students Self-reliance Class". 150 senior excellent secondary students from low-income households in targeted *poverty alleviation* areas received funds in which enabled them to complete their senior secondary education." (Company Report, 2010, p. 74).

Company	Exemplary Quotes
Statement on Engagement in a Moderate Cross-Sector Partnership Approach	
AXA	"Two avenues that we must continue to explore are the transfer of these risks to the financial markets and the implementation of *public-private partnerships.*" (Company Report, 2010, p. 43). "For example, AXA Business Services in India selected the idea of a partnership project with several NGOs to promote education for children from the *ghettos* of Pune and Bangalore." (Company Report, 2010, p. 72). "In 2010, AXA stepped up this effort by joining forces with CARE (Cooperative for Assistance and Relief Everywhere, Inc.), an internationally renowned non-governmental organization that *fights poverty* around the globe. After completing a rigorous assessment process, AXA chose this partnership as the ideal way to develop ambitious actions tied in with our *core business.*" (Company Report, 2010, p. 88).
Assicurazioni Generali	"In Ecuador, the Group *supports* FASINARM, a private non-profit foundation that provides educational assistance to children and young people suffering from *poverty* and mental disability right the way through to their integration into the job market." (Company Report, 2010, p. 111). "The Group has a long-term association with SOS-Kinderdorf, an international organization that cares for orphaned, poor and *disadvantaged children*. Each regional department "adopts" a village and provides direct aid to the children, sometimes with the personal involvement of members of staff. In 2010, the company also donated some disused PCs to the village of Pinkafeld in Burgenland." (Company Report, 2010, p. 115).
Hyundai Motor	"*Make Poverty History* is supported by the Hyundai Motor Company globally, a partnership in place since 2007." (Company Website, 2011). "HMC Joins the '2010 White Band Campaign' for *Global Eradication of Poverty*. Poverty continues to be a serious problem in many regions, and there still many who do not have enough to eat. The UN designated October 17, as the International Day of Poverty Eradication. Companies and individuals from more than 100 countries participated in this initiative by wearing a white band with 'End Poverty' on it. Many HMC employees of major overseas operation sites including the U.S., Germany, China, India and the Czech Republic have been participating in the 'White Band Campaign' since 2009. All employees worldwide were encouraged to wear the symbolic white bands." (Company Report, 2011, p. 73). "In addition, a number of 'Regional Support Network' projects that focus on the provision of comprehensive support to beneficiaries in *partnership* with regional social welfare centers were also approved." (Company Report, 2010, p. 71).
Statement on Engagement in an Extensive Cross-Sector Partnership Approach	
Chevron	"With our 40-year history in Angola, we wanted to help. But we knew we could never tackle such a project by ourselves. We turned to NGOs with expertise in agriculture, finance and education -- even seed multiplication and goat raising. We worked with international banks and development agencies, the government of Angola and, most important, with communities and rural villagers themselves. One initiative alone, aimed at reviving the nation's small farms, will have helped nearly 900,000 Angolans by the end of this year [2004]. That's about 8 percent of the entire population. The point is, success will come and will only come through new *partnerships* and coalitions that combine our separate strengths. [...] Let's join our hearts and minds in a new, shared resolve to take on the great challenge of global poverty. [...] Acting alone, no single entity or organization can *fight poverty* successfully. We need to tie companies, governments, communities and non-governmental organizations together in ways that leverage our efforts across nations and regions" (News Statement by David J. O'Reilly, Chairman and CEO, 2004 accessed

Company	Exemplary Quotes
	from Company Website, 2011). […] "Partnering for Shared Progress: We believe that business and society are interdependent. This belief drives our commitment to partnership to create mutual benefit, or shared progress. At Chevron, *partnership* is a value that we honor every day, wherever we operate, from our business to our social investments." (Company Report, 2010, p. 3). "Partnering for Sustainable Economic Growth: Following the Angola Partnership Initiative model, we are working to promote robust micro, small and medium-size businesses outside the oil industry. For example, we support the Luanda Business Incubator, a program to strengthen the operational and technical capabilities of service providers. The program has trained more than 200 entrepreneurs in Angola in business planning and helped create 143 new jobs in 2010." (Company Report, 2010, p. 11). "'Business and *community partnerships* that emphasize economic progress can help set countries on a better course.' (Ceo John S. Watson, 2011). Our investments in communities — developed in partnership with those communities — also are investments in the long-term success of our company. This approach delivers mutual benefit and shared progress. In 2010, we invested $197 million in our communities, more than twice the amount we invested in 2006." (Company Report, 2010, p. 3).
Exxon Mobil	"Developing a global *partnership for development* by supporting the EITI principles and the Universal Declaration of Human Rights, and by partnering with various organizations to further our signature initiatives. This year, we worked with the Vital Voices Global Partnership to launch the *Africa Businesswomen's Network program*, which builds and supports a network of businesswomen's organizations in Cameroon, Ghana, Kenya, Nigeria, South Africa, and Uganda. These organizations have already provided 4100 women with *business training, advocacy support, business-to-business initiatives, and capacity building*. Over the coming year, we will expand this program to Latin America and support the work of the Middle East and North Africa Businesswomen's Network. We also plan to align the work of the Businesswomen's Networks with our own supplier development initiatives, providing the training and resources to equip network members to be competitive as suppliers of ExxonMobil. […] Through our partnership with Vital Voices, we are providing training to women entrepreneurs and advocates in Africa to implement projects to help reshape laws and policies that strengthen the financial, physical, and legal well-being of women, ultimately fostering a business-enabling environment. […] We will work to engage with those who take an interest in our business and participate in constructive, progress-oriented partnerships that address global challenges and help societies gain sustainable benefits from our presence. […] Our commitment in every host nation and community where we operate is to help develop human, social, and economic capacity in a way that benefits people, communities, and our business over the long term. Achieving this is best accomplished by forming partnerships with community leaders and organizations to deliver sustainable *societal and business benefits*. […] We attempt to make strategic community investments that are aligned with global and social priorities as well as our business strengths and goals. By focusing a large part of our spending on a few major challenges of significance in the regions where we operate, we seek to enable more impactful programs; develop stronger, sustainable partnerships with governments, NGOs, and development agencies; and secure the direct engagement of our senior management and employees." (Company Website, 2011). "In partnership with Ashoka's Changemakers and the International Center for

Company	Exemplary Quotes
	Research on Women (ICRW), we implemented the Women/ Tools / Technology: Building Opportunities and Economic Power Challenge that identified 268 transformative solutions from 67 countries to promote women's economic advancement through technology. At the 2010 Clinton Global Initiative annual meeting, we announced $1 million in grants to support the expansion of high-impact, sustainable technology-driven business models identified through the challenge." (Company Report, 2010, p. 41).
Samsung Electronics	"The project is a part of KOICA's *Public Private Partnership* initiative launched in 2010 to find solutions to *poverty* in Africa. It combines the strengths of three organizations – Samsung Electronics' technology and local infrastructure, UNESCO's expertise and KOICA's finances and administrative support. Samsung Electronics employees in Korea donated KRW250 million to this cause and the company matched this, raising a total of KRW500 million." (Company Website, 2011). "Improving lives of underprivileged children is a focal point of our philanthropic endeavors in the region. We are partnering with orphanages and pediatric centers in impoverished neighborhoods to offer various support to orphans and children with cancer." (Company Report, 2010, p. 66). "Samsung Electronics' global citizenship programs reflect the needs of diverse communities around the world through *partnerships* with local organizations while domestic philanthropic endeavors focus on supporting the youth in realizing their dreams, assisting children of *low-income families*, and pursuing projects that are aligned with our business strategy." (Company Report, 2010, p. 59)."With a mission to build a prosperous future for African youth, we launched the Samsung Real Dreams program in partnership with the International Youth Foundation (IYF), which has a global network of NGOs. The program aims to increase economic activity in Africa by promoting job skills and preparing youth for successful, long-term careers." (Company Report, 2010, p. 60). "Our domestic citizenship activities focus on three areas - supporting the youth in realizing their dreams, assisting children of low-income families, and pursuing projects that are aligned with our business strategy. We are rallying all our philanthropic resources, such as stakeholder partnership, to contribute to our community in unique and powerful ways." (Company Report, 2010, p. 65). "In partnership with local NGOs and other major stakeholders, we are working to be part of the solution in solving challenges facing society by identifying the *needs of communities.*" (Company Report, 2010, p. 66).

Table 3-7: Exemplary Company Quotes on Cross-Sector Partnerships

3.2.9 Combination of Business Strategies

Concerning the *combination of business strategies* – which implies the joint combination of two or more of the four discussed business strategies at once within one project – the results are rather limited. Six possible business strategy combinations can be distinguished, being CP with Microfinance, CP with BOP, CP with SE, Microfinance with BOP, Microfinance with SE and BOP with SE. As revealed in Figure 3-17, only 11 out of 100 firms combine the presented business strategies. And hereby, 9 companies employ one combination possibility while 1 company possesses three combinations and another company engages in all six possible combinations (see Figure 3-18). Even though examples of all six combination possibilities could be obtained, it seems as if a combination of CP and Microfinance is by far the most popular combination with 8 companies undertaking it, being followed by CP and BOP as well as Microfinance and SE which are combined by 3 companies each. Exemplary quotes on the combination of business strategies can be obtained in Table 3-8.

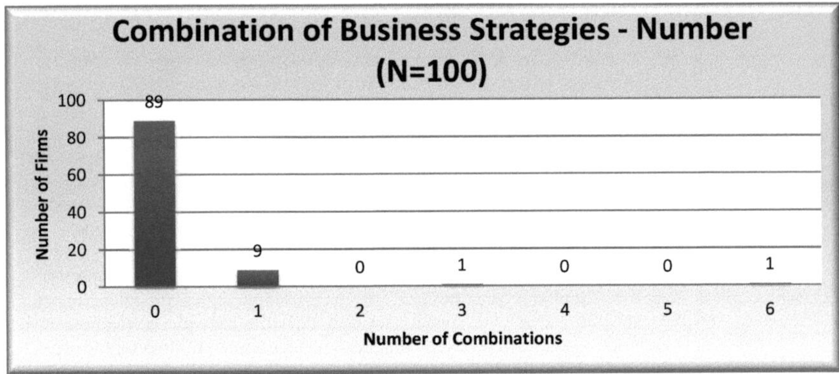

Figure 3-17: Company Overview Diagram on the Number of Combined Business Strategies

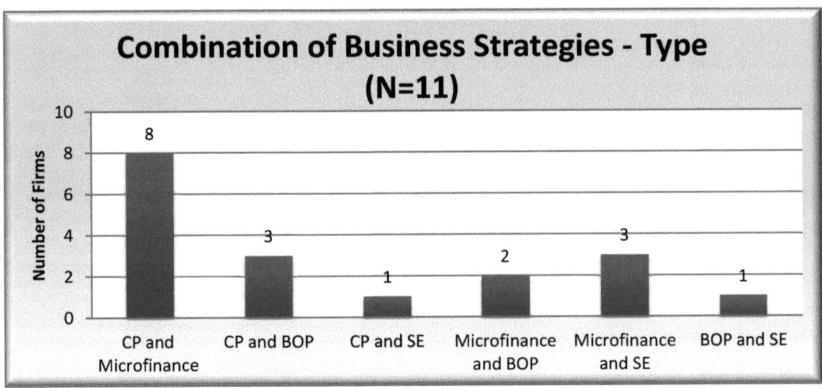

Figure 3-18: Company Overview Diagram on the Type of Combined Business Strategies

Company	Exemplary Quotes
General Electrics	"'AWF Development Debt' invests in companies that develop products and services for poor people in emerging countries, thereby contributing to poverty reduction. [...] The AWF Development Debt is an RI *fund.* [...] this fund invests in companies which do business by developing products and services for poor people in emerging countries, thereby contributing to *poverty* reduction through a '*Bottom of the Pyramid (BOP)*' approach. The fixed income fund also seeks to support *micro-credit organizations.*" (Company Website, 2011).
Carrefour	"The association *MicroFinance* Sans Frontières sends *volunteers* to microfinance institutions in emerging countries so that they can benefit from the banking skills and experience of the Group's employees and retirees." (Company Report, 2010, p. 49).

Company	Exemplary Quotes
J.P. Morgan Chase & Co.	"J.P. Morgan's Social Finance (SF) unit *provides investment and capital markets services to social enterprises,* funds, foundations, non-governmental organizations (NGOs), development financial institutions and other investors serving the base of the economic pyramid. [...] Our clients and partners include: Investors seeking social and financial returns*;* Social enterprises and impact investment funds serving the *base of the economic pyramid;* Foundations and NGOs*;* Development financial institutions *and Microfinance* networks, holding companies and investment funds." (Company Website, 2011).

Table 3-8: Exemplary Company Quotes on the Combination of Business Strategies

3.2.10 Combination of Cross-Sector Partnerships and Business Strategies

With respect to the *combination of the business strategies with Cross-Sector Partnerships* (CSPS) – which implies the undertaking of a partnering approach together with at least one of the four discussed business strategies at once within one project – 30 firms combine these approaches (see Figure 3-19). Hereby, four possible combinations can be distinguished, namely CSP with CP; CSP with Microfinance; CSP with BOP; and CSP with SE. Out of the 30 engaged firms, 23 undertake only one combination, whereas 6 companies combine two approaches and 1 actually unifies all four strategies with CSPs. Moreover, as displayed in Figure 3-20 and Table 3-9, once again examples for all combination possibilities can be obtained with CSP and CP being the most common combination with 19 active companies, followed by CSP and Microfinance with 13 companies, CSP and SE with 5 companies and CSP and BOP with 2 companies.

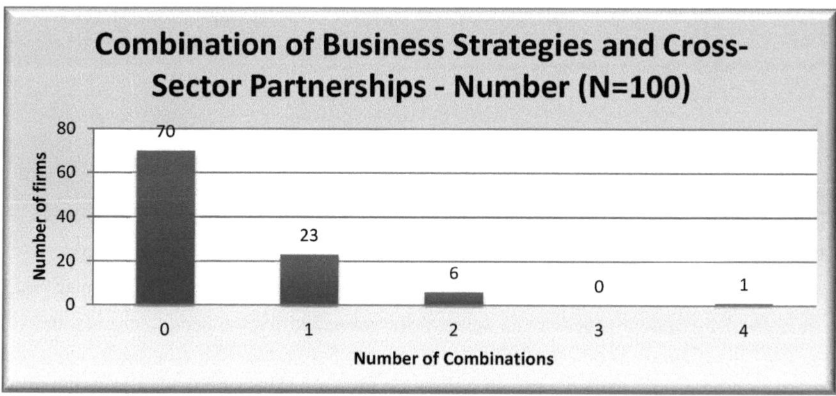

Figure 3-19: Company Overview Diagram on the Number of Combined Business Strategies and Cross-Sector Partnerships

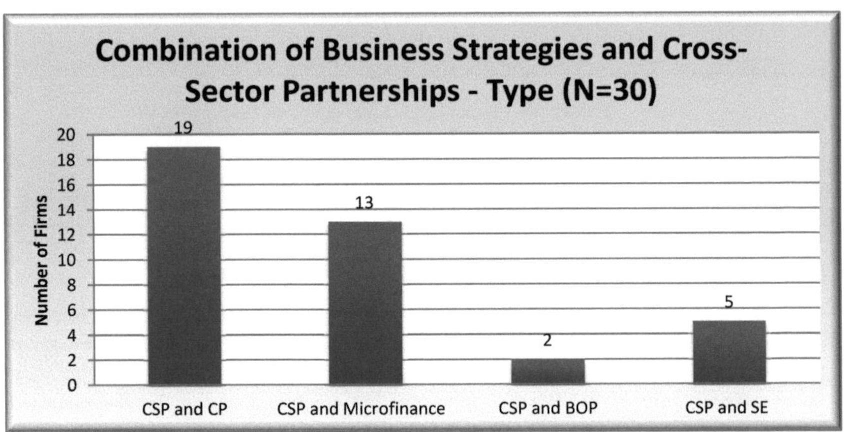

Figure 3-20: Company Overview Diagram on the Type of Combined Business Strategies with Cross-Sector Partnerships

Company	Exemplary Quotes
General Electrics	"To succeed as a global business, we need to be a part of building these societies where we operate. We do this through the products and services we create, the way we work with employees, customers, suppliers and investors, the public policies we advocate and the *philanthropic partnerships* we support." (Company Report, 2010, p. 1).
Daimler	"Together with the *non-profit partner* CARE, Daimler Financial Services is supporting with an annual global Christmas donation integrated *micro-financing* programs to provide *poverty-fighting projects.*" (Company Website, 2011).
Munich Re Group	"Munich Re helps combat *poverty* by offering *micro-insurance* cover in developing and emerging countries. In many parts of the world, Munich Re is helping develop new solutions by entering into *public-private partnerships* with suitable stakeholders." (Company Report, 2010, p. 62)

Table 3-9: Exemplary Company Quotes on the Combination of Cross-Sector Partnerships and Business Strategies

3.2.11 Overall Poverty Alleviation Portfolio Approach Classification

As outlined in the methodology, the overall *Poverty Alleviation Portfolio Approach* per firm was calculated by adding together the seven prior sub-classification scores, namely the CP Score, the Microfinance Score, the BOP Score, the SE Score, the CSP Approach Score, the Business Strategy Combination Score as well as the Business Strategy and CSP Combination Score. Figure 3-21 presents an overview of the resulting classifications of the 100 firms. Furthermore, the frequency of the resulting total scores and classifications is provided in Figure 5-2 and the scatter plot for the classification scores is additionally visualized in the Figure 5-3 of Appendix G.

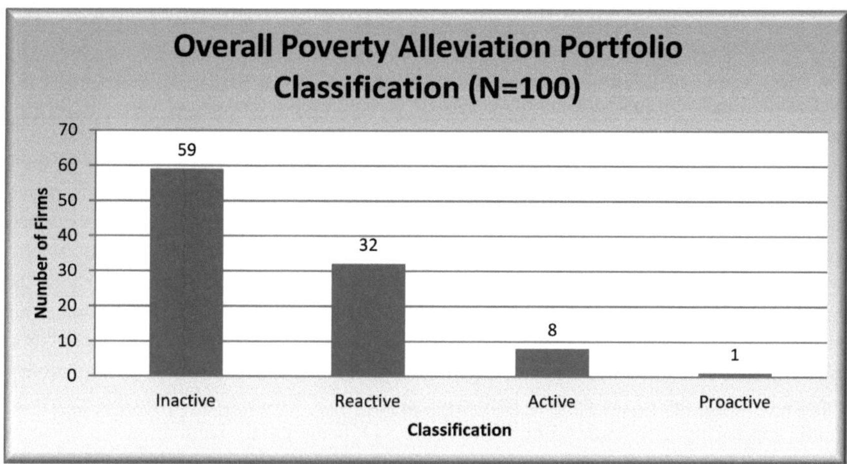

Figure 3-21: Overall Poverty Alleviation Portfolio Classification

The results thus indicate that on the whole MNCs do not seem to regard poverty alleviation as a top priority issue on their agendas. Out of the 100 examined firms, only 1 company, J.P. Morgan, engages in a proactive Poverty Alleviation Portfolio Approach with a total score of 18 out of 19. In turn, only 8 firms classify as active with 7 of them having a score of 10 and 1 a score of 11. Hence, no companies possess a score of 12, 13, 14, 15, 16, 17 or 19. Sequentially, 32 firms pursue a reactive Poverty Alleviation Portfolio Approach with 8 firms having a total score of 5; 10 firms of 6; 5 firms of 7; 7 firms of 8 and 2 firms of 9 respectively. Lastly, 59 firms, and thus being the majority of the investigated 100 firms, fall under an inactive poverty alleviation approach classification. Out of these 59 companies, 3 firms possess a total score of 0; 13 firms of 1; 15 firms of 2; 13 firms of 3 and 15 firms obtain a score of 4. To further illustrate these findings, the scatter plot in Figure 5-3 of Appendix G displays the linear function based on the distinct classification scores. Moreover, Figure 5-4 of Appendix G entails the overall classification score per company. These figures reveal that the slope of the trend line steadily decreases, the larger the score value becomes. Hence, the higher the possible obtained classification score, the less engaged are the firms in poverty-related activities.

3.3 Discussion

This section discusses the findings of the practical part. Moreover, it also aims at providing first explanations for the different Poverty Alleviation Portfolio Approaches being employed among companies. Ultimately, the findings will be illustrated through the provision of four exemplary mini-cases.

3.3.1 Discussion of the Findings

Overall, it can be stated that the majority of the sample firms appear to at least partially place the issue of poverty on their (CSR) agendas. However, in certain cases, empty phrases are detectable where companies stress the importance of poverty alleviation without actually engaging in initiatives toward tackling the issue. Hence, window-dressing might be an issue that could be further researched.

When examining the initiatives undertaken by MNCs the results tend to indicate that the pursued approaches are rather limited with the majority of the firms ultimately engaging in inactive or reactive Poverty Alleviation Portfolio Approaches while active or proactive poverty alleviation strategies are less often pursued. More concretely, it could be observed with regard to the discussed poverty strategies, that the more proactive the classification, the proportionally less engaged the firms became. Accordingly, 93 firms engaged in CP and 31 in Microfinance but solely 13 each in BOP and SE. Regarding the distinct variations, on the whole, the number of firms per variation of a strategy was more or less equal which implies that some firms actually do undertake the 'more advanced' variations. In terms of CSP, firms were relatively engaged with 73 companies participating in partnerships, accompanied by a relatively comparable allocation of the distinct partnering approaches. Besides, simultaneous combinations of approaches were rather limited with only 11 firms jointly combining the business strategies and 30 firms employing initiatives via a partnering approach. To summarize, on the whole, the overall engagement in extensive Poverty Alleviation Portfolio Approaches is very restricted. However, what is remarkable though is the fact that examples for all examined approaches – including all the variations and all investigated combination possibilities – could be obtained.

All in all, it can be concluded that companies, if engaged in poverty alleviation, primarily pursue the 'easier' and less effective approaches in addressing poverty. The more proactive the classified approach, the less engaged the firms are with only one firm, J. P. Morgan, pursuing a proactive Poverty Alleviation Portfolio Approach. As a consequence, it should be highlighted that room for improvement exists towards more proactive portfolio approaches on poverty alleviation. These results are in accordance with Kolk and van Tulder's prior findings on company reporting which state that "... frontrunner MNCs turn out to be not very outspoken, especially not on those issue that have the largest potential to alleviate poverty" (2006, p. 789) and that "...MNCs do not yet reach their full potential in offering these tools to the poor." (2006, p. 797).

3.3.2 Exploratory Analysis of the Findings

In this part, a first possible causal explanation for differences with regard to the companies' Poverty Alleviation Portfolio Approaches will be provided. Hereby, the proposed variables for the subsequent analysis are company sector, country or region of origin as well as company narrative of poverty, according to the explanation from the methodology section. A visualization of the ultimate classifications and the respective scatter plots of these variables can be found in the classification Figures 5-5 until 5-12 of Appendix G.

The above-suggested variables are of course not sufficient to fully explain a firm's selected Poverty Alleviation Portfolio Approach. There are additional possible variables to be tested in further research, which emerged from the additional literature review but were not considered in this study due to feasibility reasons. These include, for instance, firm age, company size, subsidiary influence, managerial values and the level of firm diversification (Williams and Aguilera, 2008; McWilliams and Siegel; 2001; Roberts; 1992).

Independent Variables of the Exploratory Analysis: Region of Origin and Company Sector
With regard to the *country or region of origin*, the analysis focuses on the region due to the fact that country samples would be marginal in size. However, in proceeding discussions on the academic literature, countries and regions of origin are examined jointly. The regions compared are Europe, Asia as well as North and South America with all of these three sub-groups having relatively high sample sizes of 25 or above which makes the observations more representative. A visualization of the sample size for the respective sub-groups can be found in Figure 3-22. All in all, Figure 5-5 and 5-6 of Appendix G indicate that slight differences in outputs among the sub-groups can be obtained with the overall classification tendencies nonetheless being comparable. More precisely, North and South America seems to score highest overall, followed by Europe and lastly Asia. As a result, findings appear to be slightly contingent to the region of origin.

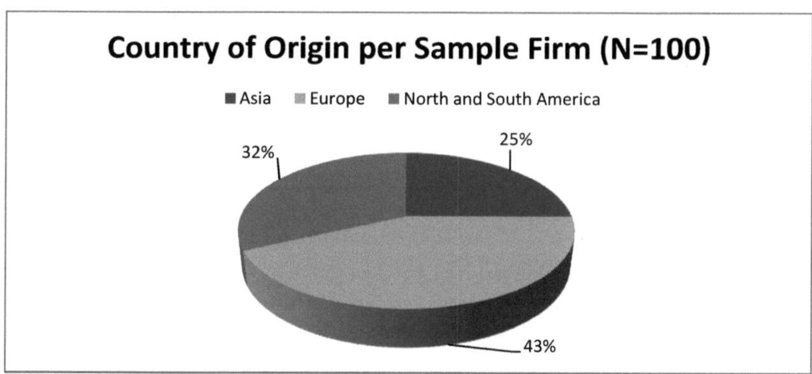

Figure 3-22: Country of Origin per Sample Firm

Next, considering the *company sector*, eight distinct industry classifications were compared, as displayed in Figure 3-23. The selection of such a number of sub-groups obviously reduces the sample size per sub-group down to five at the lowest. Such small sample sizes overall restrict generalizations to the population, with the possible exception of oil and gas as well as financials due to their industries' larger sample sizes. Thus, only inferences on trends can be made. All in all, Figure 5-7 and 5-8 of Appendix G indicate that the slopes of the trend lines diverge per sector, with the technology and the financial sectors classifying as most engaged, the utility firms as least engaged, with the remaining sectors falling in between. The differences among the sectors appear to be stronger than the differences in classification among the regions of origin, which could also simply be a result of the small sample sizes per sub-group, which makes extreme outcomes more likely. Consequently, company sectors might influence a firm's Poverty Alleviation Portfolio Approach.

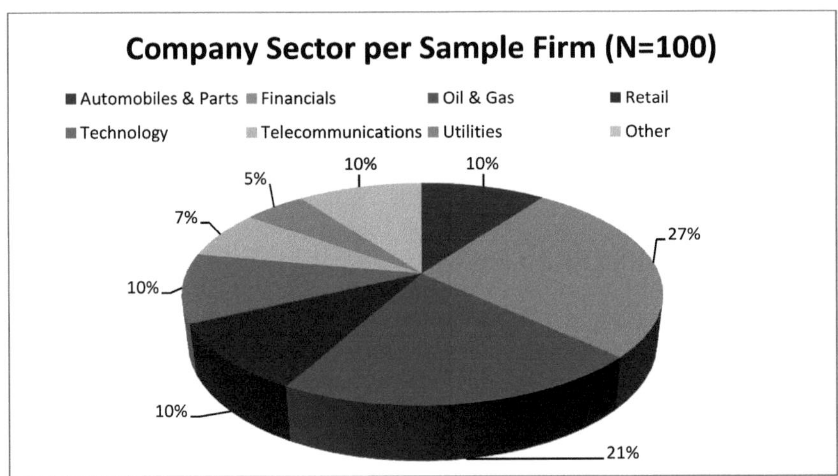

Figure 3-23: Company Sector per Sample Firm

Overall, these findings indicate that a firm's Poverty Alleviation Portfolio Approach appears to be at least partially contingent on a company's region of origin as well as its sector. This is in accordance with van Tulder's (2010) results which stress that companies' poverty alleviation approaches differ among regions and sectors. Along these lines, McWilliams and Siegel (2001, p. 126) obtain a comparable result by concluding that "... the provision of CSR will vary across industries, products and firms" whereas Dawkins and Lewis (2003, p. 189) state "... the impact of a company's corporate responsibility program will vary according to the industry sector." Furthermore, Williams and Aguilera (2008, p. 22) claim that "... the country of origin has a significant effect [...] on what type of information is most likely to be covered in these MNCs'[...] reports." Similarly, Wanderly et al. (2008, p. 369) emphasize "... that both country of origin and industry sector have a significant influence over CSR information disclosure on the web." However, Wanderly et al. also find that differences in outcome are stronger for the country of origin than for the sector whereas the findings in this research indicate the opposite. One possible explanation for the ambiguous results may be the fact that the former study focuses on emerging countries whereas the sample firms in this research originate primarily from developed countries. Possible factors which might impact the relationship between country of origin or sector and the pursued poverty alleviation approach include the political cultural, the freedom of press and the overall CSR notion (Wanderly et al., 2008; van Tulder and van der Zwart, 2006). Moreover, stakeholder expectations, the business environment, the government-business relationship as well as the legal or institutional situation might also be highly influential (Williams and Aguilera, 2008; Chapple and Moon, 2005).

Mediating Variables of the Exploratory Analysis: Issue Prioritization and Issue Framing

Based on the methodology described in the prior data analysis section, the *issue prioritization of poverty* per firm has been derived. The results in Figure 3-24 reveal that 16 out of 100 firms tend to at least partially prioritize poverty. However, as poverty was conceptualized broadly, the findings reveal that poverty has not once been directly stated as focus theme. Instead, 12 firms list as focus area community development or engagement in communities whereas 3 companies emphasize social inclusion and 1 company highlights rural development. All these themes partially, but not entirely, encompass the issue of poverty, with poverty not necessarily having to be the only theme addressed. This might also explain why no substantial difference between the two sub-groups was found, as displayed in Figure 5-9 and 5-10 of Appendix G.

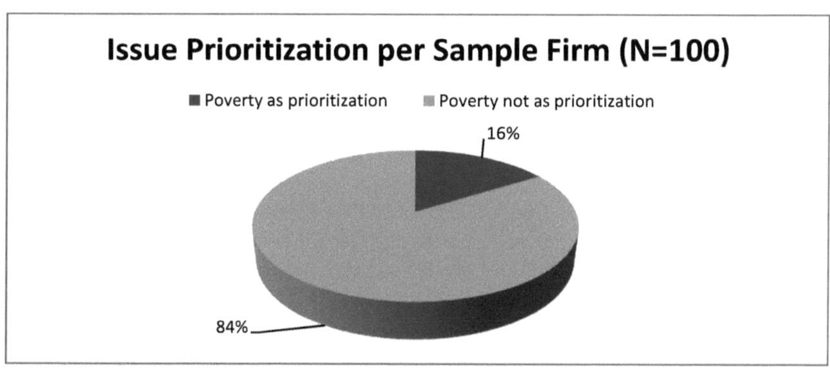

Figure 3-24: Issue Prioritization per Sample Firm

In turn, a visualization of the distinct *issue framing* categories based on the poverty statements – as described in the methodology section above – can be obtained from Figure 3-25. The findings reveal that 51 of the 100 sample firms do not possess a poverty alleviation statement and hence, do not frame the issue of poverty. Moreover, of those which frame the issue, the overwhelming majority, namely 37 firms, frame poverty in a positive manner, whereas solely 7 and 5 firms respectively frame poverty in neutral or negative terms. When comparing the four sub-groups (see Figure 5-11 and 5-12 of Appendix G), the findings indicate differences with regard to the employed Poverty Alleviation Portfolio Approach among these sub-groups. Specifically, those firms, which frame poverty, tend to outperform those, which do not frame the issue. Among the companies that frame poverty, however, the firms, that regard poverty in neutral or negative terms, appear to score higher on their Poverty Alleviation Portfolio Approaches than those that frame poverty positively. This finding is counter-intuitive and might be a result of the small sample sizes, in particular with regard to the neutral and negative farming sub-groups. Nonetheless, it could also be reasoned that the more firms perceive the issue of poverty as a threat, the more engaged firms will become in poverty-alleviating activities in order to minimize the possible negative effect of poverty on the company and society.

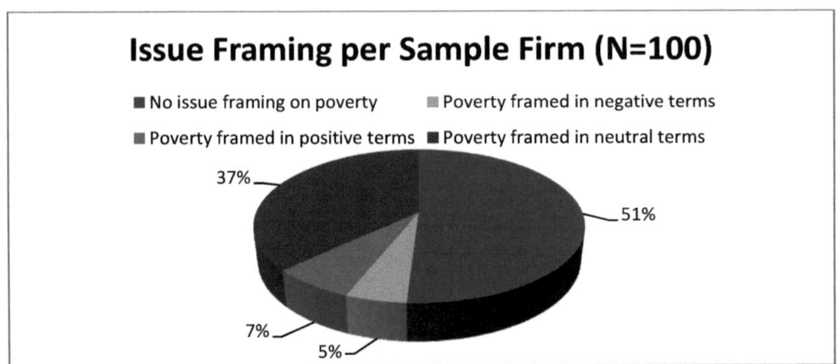

Figure 3-25: Issue Framing per sample Firm

The literature review on the relationship between CSR issue framing as well as CSR issue prioritization and CSR reporting or engagement has found that differences among sub-groups exist, but does not further specify the extent of these differences or the directions of the relationships. As an example, Williams and Aguilera (2008, p. 21) remark the following in their study: "Interesting differences emerge in what issues are emphasized in companies' sustainability reports and how these issues are framed." This research as well as another research by Chapple and Moon (2005) and Dawkins and Lewis (2003) further tend to analyze these findings according to a firm's country of origin and the company sector. For instance, Chapple and Moon (2005, p. 418) remark that differences in CSR engagement and reporting could be obtained between Europe and the US with regard to the "... issues engaged in [...] which reflect the salient national business-society agenda." This is in line with Wanderly et al. (2008, p. 372) who stresses that "... companies focus on issues that are relevant to them..." and their social context. These observations support the position of a company narrative on poverty as a mediating variable between country of origin and Poverty Alleviation Portfolio Approach. Hence, potential reasons for the differences in issue prioritization and issue framing among sub-groups could be the external context, such as the local prioritization of matters in a firm's direct political, economic or cultural environment. With regard to the industry, the institutional and legal environment as well as stakeholder demands or competitive benchmarks might also impact the issue framing and prioritization. Issue framing and prioritization can in turn provide the underlying basis for general as well as CSR specific corporate decisions and therefore naturally impact the engagement level of a company with regard to a certain issue, such as poverty. A more in-depth study of these relationships and their contingent factors would be a reasonable suggestion for further research

Observations from the Exploratory Analysis

Further analysis of potential causal explanations of differences with regard to companies' Poverty Alleviation Portfolio Approaches has highlighted company sector as well as country or region of origin and company narrative on poverty as possible influential variables. More specifically, comparisons in classifications among sub-groups were most significant for company sector and issue framing, followed by region of origin and ultimately least differences could be observed between the issue prioritization sub-groups. However, due to the overall small sample sizes per sub-group and the lack of empirical testing, purely indications and no validated statistical inferences can be drawn from the findings. A further literature review in this regard has revealed that all four examined variables could potentially influence a companies' Poverty Alleviation Portfolio Approach as dependent variables, with company sector and region of origin most likely serving as independent variable and issue framing and issue prioritization being studied as mediating variables. Further research could build upon these first indications by empirically testing the concrete relationships between the proposed variables and a firm's Poverty Alleviation Portfolio Approach by comparing the classifications scores among the sub-groups.

3.3.3 Illustration of the Findings

An in-depth exploration of the potential explanatory variables per company would have to be conducted via in-depth case studies, which exceed the scope of this thesis. However, in order to provide exemplary analyses, which consider a firm's specific context, four mini-case studies were prepared. Each mini-case study in turn examines one possible classification approach. These cases serve to illustrate, investigate and elaborate on the findings obtained per company and can be found in the proceeding Boxes 3-1 to 3-4.

Overall, the illustrated mini-cases indicate that firms' poverty-related initiatives can be highly diverse in its concrete implementation manner, being contingent on a company's concrete context and strategy. Accordingly, in-depth case studies might facilitate exploratory analyses of the underlying reasons for firm-specific classifications. Moreover, the proposed explanatory variables appear to be at least partially influential for the undertaken approaches. In general, the internal variables, namely issue framing and issue prioritization on poverty, seem to be more decisive than the external variables of company sector and region of origin. Moreover, a further key variable for firms' strategic choices within the field of poverty alleviation might not solely be the firms' narratives on poverty but their narratives on CSR in general. Hence, an analysis on the manner in which companies prioritize and frame CSR activities could contribute to an enhanced understanding of firms' ultimate Poverty Alleviation Portfolio classification choices.

Mini-Case 1: Berkshire Hathaway

An Exemplary Analysis of an Inactive Poverty Alleviation Portfolio Approach

Company Background:

Berkshire Hathaway is ranked as 19th largest global company according to the global Fortune 500 firms ranking of 2011 with annual revenues of $136,185 million and a profit of $12,967 million in 2010 (Fortune website, 2011). The company's current chairman and CEO is 'business legend' Warren Buffet who is regarded as one of the wealthiest and most successful managers (Fortune website, 2011). The company is headquartered in Nebraska, the *United States* and thus is considered as a *North and South American* firm with regard to the region of origin (Company website, 2011). Regarding the sector, according to the Industry Qualification Benchmark (2011), Berkshire Hathaway classifies as *financial firm*. However, to be concrete, the company is actually a conglomerate which owns around 70 firms, primarily financials but also companies from other sectors, such as textile, jewelry or oil and gas (Company website, 2011).

CSR Reporting:

With regard to its social responsibility reporting, Berkshire Hathaway is besides American Group International the only global 100 company which makes *no reference to any CSR-related matters or activities* undertaken on its website. Moreover, no CSR or any other non-financial report could be obtained from the company's website. Accordingly, a company statement on the issue of poverty is not provided either. This implies that due to a lack of focus on poverty or any other CSR issue and due to no issue framing of poverty, *no company narrative on poverty* can be specified. Overall, it can be remarked that the company's website – as well as the company's general orientation – appears to be highly shareholder focused, with plentiful information being available on stock prices, financial performance and investment backgrounds. To illustrate, Berkshire Hathaway states as its long-term goal: "... to maximize Berkshire's average annual rate of gain in intrinsic business value on a per-share basis." (Company website, 2011).

Engagement in Poverty Alleviation Activities:

As result of the *lack of CSR reporting* on the website and the absence of a CSR report, no information on any engagement into poverty-alleviating activities could be obtained. Further internet research has revealed that Berkshire Hathaway is widely criticized for its non-engagement in CSR matters. Hence, it can be presumed that Berkshire Hathaway does not pursue any poverty-related business approaches.

Poverty Alleviation Portfolio Classification:

As a result of its non-engagement in any business initiatives with regard poverty, Berkshire Hathaway clearly classifies as undertaking an *inactive Poverty Alleviation Portfolio Approach* with its total score being equal to 0.

Comparison and Analysis of Results:

With regard to the proposed indirect variables for further examination, this mini-case study of Berkshire Hathaway is contradictory to the prior findings, that generally firms, which originate from the United States/America and which operate in the financial sector, might be more likely to pursue an active or proactive poverty alleviation portfolio approach. In the case of Berkshire Hathaway possible explanatory variables for the company's inactive classification might be foremost *internal* reasons. Hence, the firm does not seem to focus on CSR matters overall but prioritizes shareholder demands and due to this general inactive notion on CSR, engagement in poverty-related activities is unlikely to be expected.

Box 3-1: Exemplary Mini-Case for an Inactive Classification – Berkshire Hathaway

Mini-Case 2: Chevron

An Exemplary Analysis of a Reactive Poverty Alleviation Portfolio Approach

Company Background:

Chevron ranks as 10th largest global firm based on the global Fortune 500 firms ranking of 2011 with annual revenues of $196,337 million and a profit of $19,024 million in 2010 (Fortune website, 2011). The company possesses two headquarters in the *United States* and in Brazil and therefore falls under *North and South America* as region of origin. Moreover, Chevron classifies as an *oil and gas* company with regard to its sector (Company website, 2011).

CSR Reporting:

Chevron does provide a *CSR section* as well as a *CSR report* on its website. The company's slogan is 'human energy' and this humanity reappears as central theme throughout the firm's *extensive CSR reporting* sections. Poverty is *not* framed as a *CSR prioritization* issue and instead the Chevron appears to focus foremost on energy and environment-related initiatives in addition to health and safety as well as human rights. Besides, Chevron *frames* the issue of poverty in *positive terms* by explicitly including a CEO statement on poverty in which the company describes in depth how it could contribute to tackling poverty and in which the CEO frames poverty as follow: "Poverty is of deep concern to me. I believe the issue of poverty and the meeting of basic human needs is the defining challenge of the 21st century." (Company website, 2011). However, as a contradicting note, it has to be remarked that Chevron does not appear to be as engaged in CSR matters in practice as claimed in its statements. To be concrete, Chevron has been highly publicly criticized for its irresponsible environmental behavior and ignorance of stakeholder demands. Hence, it retrieved the *Public Eye Award* for irresponsible behavior (Tagesschau website, 2011).

Engagement in Poverty Alleviation Activities:

Regarding the undertaken poverty initiatives, Chevron engages in *broad Microfinance via an extensive partnering approach*. In this regard, initiatives stress the importance of partnering and empowerment with the following underlying reasoning on Microfinance being provided: "Microfinance is not charity. It's about building capability and empowerment, and it places the responsibility for success on the participants. [...] Fighting poverty should not be something we do to people it should be something we do with them." (Company website, 2011).

Poverty Alleviation Portfolio Classification:

To summarize, Chevron does not undertake any CP, BOP or SE activities. Moreover, no combination of business strategies could be obtained. Due to its engagement in broad Microfinance (score of 2), its pursuing of an extensive partnering approach (score of 3) as well as its employment of Microfinance via a partnering approach (score of 1), Chevron obtained a total score of 6. As a result, Chevron classified as following a *reactive Poverty Alleviation Portfolio Approach*.

Comparison and Analysis of Results:

Underlying reasons for Chevron's classification are unlikely to be external factors as American firms were identified as most active region and as the oil and gas industry also appears to be moderately engaged in poverty. In turn, internal motives might be more likely to explain he classification approach, such as the positive issue framing approach. In particular, the results indicates that corporate reporting on CSR in general and on poverty in particular is extensive, as the firm apparently wants to be seen as highly committed to sustainable development goals. In contrast, the prior analysis clearly indicates a case of green-washing whereas Chevron presents itself as highly CSR-focused firm while not actually fulfilling this image in practice. Hence, a more in-depth analysis of Chevron's corporate communication strategy on CSR could be helpful in better understanding the possible influencing factors for the firm's Poverty Alleviation Portfolio Approach.

Box 3-2: Exemplary Mini-Case for a Reactive Classification – Chevron

Mini-Case 3: Crédit Agricole

An Exemplary Analysis of an Active Poverty Alleviation Portfolio Approach

Company Background:

According to the 2011 Fortune ranking, Crédit Agricole is considered the 43rd largest global firm with revenue of $105,003 million and profit of $1,672 million in 2010 (Fortune website, 2011). The firm is headquartered in Paris, *France*, and therefore classifies – as only out of the four mini-cases – not as an American but as a *European firm* on region of origin (Fortune website, 2011). Crédit Agricole is a banking group and consequently is regarded as *financial firm*.

CSR Reporting:

For Crédit Agricole both a *CSR report* and a *CSR section on the website* could be accessed. In general, the firm's company reporting is relatively extensive with explicit commitments to Sustainable Development being highlighted on the company website. It nonetheless remains questionable whether such an extensive CSR notion is penetrated in practice or whether the company engages in *green-washing activities*. This uncertainty results from the fact that Crédit Agricole has receives fines in the past for irresponsible Corporate Governance practices. Concerning the precise company narrative on poverty, it first of all has to be remarked that poverty does *not* appear to be *prioritized* in the firm`s CSR activities. In contrast, emphasis is placed on the company's website on stakeholder groups other than the poor, namely the environment, customers or employees. With regard to the *issue framing approach*, Crédit Agricole clearly presents poverty in *positive* terms by highlighting the company's "... contribut[ion] to international development aid in areas such as combating poverty and exclusion." (Company Report 2010, p. 163).

Engagement in Poverty Alleviation Activities:

Considering the undertaken initiatives on poverty, Crédit Agricole engages in *narrow CP* by "... providing financial support ..." to external NGOs that "... aid populations at risk." (Company website, 2011). Moreover, the company pursues a *broad Microfinance* strategy by setting up an own Crédit Agricole Grameen Foundation on Microfinance which does not only offer loans but also financial assistance and which focuses on particularly targeting the poorest. This strategy also encompasses extensive Microfinance *partnerships* via NGOs (Company Report, 2010). Besides, engagement in narrow social business activities is sought which also aims at providing financial access to the poor and thus *combine Mircofinance and SE*. In this context of poverty and Microfinance, Crédit Agricole announces the following: "By supporting microcredit organisations we help to give the poorest people access to financing, enabling them to grow their businesses and lift themselves out of poverty." (Company website, 2011).

Poverty Alleviation Portfolio Classification:

To conclude, Crédit Agricole does not undertake any BOP activities. However, the firm does engage in classical CP (score of 1), broad Microfinance (score of 2) and narrow SE (score of 2). Besides, the firm also employs an extensive CSP approach (score of 3) and it possesses one business strategy as well as one business strategy-partnership combination (score of 1 each). Accordingly, based on a total score of 10 Crédit Agricole was classified as pursuing an *active Poverty Alleviation Portfolio Approach*.

Comparison and Analysis of Results:

Overall, in its poverty-related activities, Crédit Agricole highly focuses on Microfinance initiatives and combinations. This seems plausible since the firm is a financial company and therefore Microfinance is closely related to the company area of expertise. Other explanations for Crédit Agricole's classification could be the positive issue framing on poverty, which underscore the firm's believe of being able to make a contribution, or the general manner in which the company aims to present itself on CSR – maybe also due to external pressures – namely as highly engaged and committed firm.

Box 3-3: Exemplary Mini-Case for an Active Classification – Crédit Agricole

Mini-Case 4: J. P. Morgan Chase & Co.

An Exemplary Analysis of a Proactive Poverty Alleviation Portfolio Approach

Company Background:

J. P. Morgan ranks at 36th largest global firm according to the Fortune 2011 list, with a revenue $115,475 million and a profit of $17,370 million in 2010 (Fortune website, 2011). The company's name dates back to its founder 'J. P. Morgan' who was one of the most famous and influential bankers of his times (Company website, 2011). The firm's headquarter stands in New York, the *United States*, and thus it classifies as *North and South American firm* regarding the region of origin. Due to the fact that J. P. Morgan is a bank, it is placed within the *financial firms* as company sector (Company website, 2011).

CSR Reporting:

On J. P. Morgan's website, an *elaborate CSR section and CSR report* can be obtained. The company *prioritizes poverty* in terms of community development and also highlights other focus themes, such as the environment or diversity. Besides, it is highlighted that J. P. Morgan adheres to strict standards in the field of corporate governance and CSR (Company website, 2011). Besides, poverty is *framed in positive terms*. Overall, J. P. Morgan publishes the following statement which emphasizes the company's general commitment to CSR: "We believe that being profitable and doing good works for the people and the world around us aren't exclusive of each other; they are integrated goals." (Company website, 2011).

Engagement in Poverty Alleviation Activities:

To sum it up, J. P. engages in all possible approaches, their broadest variations and fullest combination possibilities and, with the exception of its partnering approach, the firm always receives the highest possible score. An illustration of the extensive, joint undertaken initiatives on poverty is the following selection of quotes: "JPMorgan Chase's *philanthropic goal* is simple – be a catalyst for meaningful, positive and sustainable change within our highest-need neighborhoods and communities across the globe. In 2010, JPMorgan demonstrated its support of organizations serving the *base of the economic pyramid* by expanding its commitment to *Grameen Foundation* [...] J. P. Morgan's Social Finance (SF) unit provides investment and capital markets services to *social enterprises*, funds, foundations, NGOs, development financial institutions and other investors serving the base of the economic pyramid. In doing so, SF achieves a *double bottom line* of social benefit and financial return. [...] Working with *best-in-class community-based partners*, our goal is to help stabilize families living in high-poverty neighborhoods and to make that stability echo through a neighborhood in a manner that improves educational and job opportunities, reduces crime and dramatically raises the community's quality of life." (Company website).

Poverty Alleviation Portfolio Classification:

To summarize, J. P. Morgan pursues *strategic CP* (score of 2), *broad Microfinance* (score of 2), *inclusive BOP* (score of 3), *broad SE* (score of 3) in addition to a *moderate CSP approach* (score of 2). Besides, *all six business strategies* as well as *all four business strategy-partnership combinations* were employed at once (score of 3 each). As a result, J. P. classifies with a total score of 18 – as only firm of the sample – as undertaking a *proactive Poverty Alleviation Portfolio Approach.*

Comparison and Analysis of Results:

All in all, J. P. Morgan's proactive classification on poverty might be partially explained by its sector and region. More concretely, the previous analysis has indicated that, on the whole, financial firms which originate from North or South America tend to be more engaged in poverty-related initiatives. Besides, the employed issue prioritization and positive issue framing on poverty could further explain J. P. Morgan's poverty alleviation approach. Ultimately, the general research on the firm's CSR reporting suggests that the company is overall highly committed to CSR matters which in turn can also contribute to the proactive classification of the firm.

Box 3-4: Exemplary Mini-Case for a Proactive Classification – J.P. Morgan Chase & Co.

A Concluding Remark

To summarize the practical section of this thesis examined the following two, previously presented research questions:

- *Which poverty-related initiatives are actually undertaken by Fortune's largest 100 MNCs?*
- *In how far do the Poverty Alleviation Portfolio Approaches by the largest 100 MNCs reach up to the theoretical potential of the private sector in effectively addressing poverty?*

With regard to the first research question it can thus be concluded that almost all companies of the sample, namely 97 out of 100 firms, undertake one or more poverty-related initiatives. Impressively, company examples for all examined approaches, as well as their variations and combination possibilities, could be obtained. The most common initiative is Corporate Philanthropy followed by Cross-Sector Partnerships and Microfinance. In contrast, activities related to the Bottom of the Pyramid or Social Entrepreneurship strategies are much less pursued. In turn, the extent of engagement in the distinct variations was on the whole similarly spread and no extreme cases could be observed. Ultimately, only a small percentage of firms combine more than one distinct business approach at once – be it the presented business strategies among themselves or in combination with Cross-Sector Partnerships.

Sequentially, considering the second research question it can be stated that the undertaken company approaches classify mostly as inactive or reactive and only in limited cases as active or proactive regarding their Poverty Alleviation Portfolio Approaches. Most of these pursued approaches have been regarded as insufficient to effectively tackle poverty in the literature review section. Hence, while most companies pursue poverty-related activities they primary choose the less 'complex' and less effective initiatives. As a result, it can be concluded that at present the overall Poverty Alleviation Portfolio Approaches of MNCs are rather limited with substantial room for improvement and thus the examined firms do not reach up to the theoretical potential of the private sector in effectively and sustainably addressing poverty.

4 Discussion and Conclusion

This chapter discusses and concludes the theoretical and practical research conducted above. Thus, first of all, the theoretical and practical research contributions are presented, being followed by an overview of the implications of the findings on academia and business operations. Afterwards the main findings of this thesis are summarized and research limitations as well as suggestions for further research are stated.

4.1 Research Contributions

This section outlines the theoretical and practical contributions of this thesis. Hereby, first the identified research gaps and shortcomings are revealed followed by an examination of how these gaps are addressed in the current study.

4.1.1 Theoretical Research Contributions

In this part the theoretical research gaps as well as the contribution of this study in addressing these gaps are presented.

Identified Research Gaps and Shortcomings

With regard to the identified theoretical research gaps, it has to be mentioned, that whereas the field of poverty and its alleviation has been widely researched, the area of 'business approaches to poverty alleviation' has only recently began to receive substantial academic attention. However, thus far results have been overall inconclusive and relatively vague. Based on the extensive analysis of the related literature in this thesis, several observations could be made.

To start, various approaches and themes discussed appear to be reoccurring. In this manner, in particular the importance of companies for tackling poverty is frequently highlighted whereas in-depth analytical analyses or empirical tests rarely follow. Hence, much is presumed but verifications of assumptions in practice are lacking. Hereby, it should also be stressed that many of the aspects discussed have been poorly defined and findings have been controversial, such as the concrete impact of business initiatives on poverty alleviation, or concepts like Social Entrepreneurship or the Bottom of the Pyramid.

Moreover, while particular business approaches on poverty reduction have been subject of much academic debate, a frequent perspective of analysis was the business itself and the implications of active engagement on business operations. In contrast, the impact of business initiatives on the wider societal context and specifically the poor has been less extensively discussed. Additionally, research focus has been predominantly set on examining distinct business approaches in isolation. Combinations or comparisons of approaches per se rarely occurred and thus far no classification of various poverty alleviation approaches could be retrieved – besides the classification by van Tulder (2010) himself, which this study builds upon. Accordingly, a placement of individual initiatives in the broader context as well as an analysis of companies' portfolios on poverty alleviation has been lacking in past research.

These identified theoretical research gaps have been highlighted in prior studies as well. As an example, Pfefferman (2001, p. 42) announced that "… little has been published about the development impact of private enterprises." In related terms, Frynas (2008, p. 275) states that: "… while private sector development initiatives can perhaps be beneficial for specific firms in terms of reputation effects or new product development, we know relatively little about their developmental benefits." Ultimately, Raufflet, Berranger and Aguilar-Platas (2008, p. 54-55) remark the following on the academic field of business and poverty:

"Poverty is still a poorly conceptualized concept in management and business in general. [...] Connecting business strategy with poverty is new to management research. [...] These approaches are still in their infancy: they lack solid conceptual definitions and rigorous models of social change. These approaches would probably gain from systematic research that would evaluate their effects on poverty alleviation."

Addressing the Research Gaps

This thesis therefore addresses the previously identified research gaps in the area of business approaches on poverty alleviation by outlining main theoretical and practical contributions of MNCs with regard to poverty eradication. More concretely, this thesis engages in a distinct approach by connecting and *jointly* analyzing previously separate research fields. On the one hand, a unified analysis of five selected business strategies on poverty alleviation, together with the theme of Cross-Sector Partnerships, was conducted. On the other hand, the potential for combination of these distinct business approaches was explored. This study thereby looked at the entire portfolio of company initiatives – in theory and in practice – and accordingly, a classifications framework of companies' Poverty Alleviation Portfolios Approaches was proposed. Since such a classification is a unique undertaking in this academic field, it should consequently enrich existing research by enabling an evaluation and comparison of company approaches on poverty alleviation in theory and in practice.

4.1.2 Practical Research Contributions

This part highlights the practical research contribution in addressing the identified practical research gaps and shortcomings.

Identified Research Gaps and Shortcomings

Concerning the practical research contribution, the same argument as above can be employed, namely that the topics addressed in this thesis have thus far not been investigated jointly in such breadth in practice. The literature review revealed that the overwhelming majority of articles solely consist of literature reviews, with primary research being rarely conducted. Furthermore, if practical studies in the field of business and poverty were undertaken, they have generally been qualitative in nature and based on a small number of case studies (van Tulder, 2010). Similarly, MNCs have seldom been employed as units of analysis, with studies mostly conducted from the perspective of SMEs as well as governmental and nongovernmental actors.

Once again, these observations on practical research gaps are also confirmed in several studies. For instance, Lyon and Bertotti (2007, p. 177) highlight that: "…at present there is a lack of robust empirical evidence to demonstrate the actual contribution of enterprises to the reduction of poverty and deprivation." Similarly, other researchers in the field of business and poverty remark the following: "One of our key observations is the lack of knowledge about best practices and the great need for more sustained research and analysis of what works and what doesn't" (UNDP, 2004, p. 2). As a final example, van Tulder (2010, p. 30) concludes that "that the business involvement in addressing the issue of poverty is far from settled, first by a lack of meaningful benchmarks, approaches and measurement tools, not by a lack of best practice cases."

Addressing the Research Gaps

As a consequence, the research in this study is unique due to the fact that it simultaneously analyzes and classifies the Poverty Alleviation Portfolios of the largest 100 MNCs by examining their websites and CSR reports. The study is thus based on a broader and therefore more representative sample; it uses a distinct approach of data collection through content analyses of corporate reporting; and it employs MNCs as unit of analysis. Besides, classifications of business initiatives based on companies' entire Poverty Alleviation Portfolios have not been conducted in a comparable manner in prior research. Consequently, a unique overview of the types, variations and combinations of the actual initiatives undertaken by MNCs on poverty alleviation is provided, which should enable a first comparison of MNCs' Poverty Alleviation Approaches in practice.

Overall, by developing classifications of various business approaches to poverty alleviation, as well as their variations and combinations, and by applying the proposed classification framework to practice, this thesis contributes significantly to the existing research field.

4.2 Research Implications

This section of the thesis will present an overview of the main research implications for academia and for business managers.

4.2.1 Theoretical Implications

As already outlined in the research contribution, this thesis has taken up a distinct approach by jointly analyzing and classifying a variety of business initiatives on poverty alleviation. It thereby enriches existing research in the field, which has primarily focused on examining one particular approach in isolation. Consequently, the analysis undertaken in this study could encourage further interdisciplinary research in the area of business and poverty which focuses on the 'bigger picture' and the connectivity among initiatives. Moreover, this study could stimulate researchers to further (re)classify and evaluate distinct business approaches on poverty alleviation – in theory and in practice – by building up on the classification framework presented in this research.

4.2.2 Practical Implications

The results of the conducted research indicate that MNCs currently do not appear to sufficiently prioritize the issue of poverty in their business operations and thereby do not fully explore their poverty alleviating potential. It must be recognized, though that this paper has analyzed company engagement in poverty alleviation approaches from a societal and not from a business perspective. However, involvement in poverty-related activities can be considered crucial from a strategic business viewpoint as well due to several reasons.

To begin, it has frequently been highlighted that stronger business involvement in societal issues, such as poverty alleviation, is a necessity, as non-involvement could threaten future business success. As Helsin and Ochoa (2008, p. 141) state it "...there is no business to be done on a dead planet." While this quote relates foremost to environmental degradation, it can nonetheless be transferred to poverty, where worsening conditions could, for instance, threaten the political and economic stability of business environments.

Next, non-involvement in activities on poverty or other issues has the potential to harm stakeholder relations and firm reputation (Sharp, 2006; Tang and Li, 2009). In contrast, the adoption of a proactive strategy – which, as the results of the study indicate, is not widely adopted yet – might benefit stakeholder relations with concerned customers, employees, communities or the government.

Furthermore, several of the presented business strategies, such as Bottom of the Pyramid, Social Entrepreneurship or Microfinance, also have the potential to provide firms with a strategic opportunity for additional growth and revenue generation (WBCSD, 2004). These business cases on poverty alleviation approaches are particularly beneficial because they represent a win-win situation for companies and for society (McFalls, 2007; Goldsmith, 2011).

And finally, engagement in pressing global issues, such as poverty, has to be regarded as a moral obligation – in particular for highly influential, global firms (Warhust, 2005). Besides the potential business benefit, managers could also simply pursue a proactive approach simply because it is the right thing to do.

Hence, as a result of the prior argumentation, stronger business engagement through a proactive poverty alleviation approach is highly recommendable and should be seriously considered for firms who want to successfully and integrally lead in the future. For those managers who choose for a more proactive strategy, this thesis can subsequently offer first guidance on which approaches to pursue in order to most effectively address the issue of poverty. Eventually, the research results of this study can also be employed to analyze or benchmark the performance of certain companies with regard to poverty alleviation.

4.3 Conclusion

This section summarizes the undertaken approach as well as the main finding of the study. Overall, this thesis examined the theoretical and actual contribution of MNCs to poverty alleviation by classifying companies' Poverty Alleviation Portfolio Approaches. Hereby, the following research questions were addressed:

- *What are the roles and potential contributions of the business sector, in general, and MNCs, in particular, towards combating poverty?*

- *In how far are the examined business strategies on poverty alleviation – being Business-as-Usual; Corporate Philanthropy; Microfinance; Bottom of the Pyramid and Social Entrepreneurship – and their variations effective in addressing poverty?*
- *In how far are Cross-Sector Partnerships effective in addressing poverty?*
- *In how far can the prevailing poverty alleviation initiatives be combined and in how far would such a joint approach increase the effectiveness of the undertakings?*
- *Which poverty-related initiatives are actually undertaken by Fortune's largest 100 MNCs?*
- *In how far do the Poverty Alleviation Portfolio Approaches by the largest 100 MNCs reach up to the theoretical potential of the private sector in effectively addressing poverty?*

In order to address these questions from a theoretical standpoint, first, a brief introduction to the issue of poverty and its alleviation approach was provided, followed by an examination of the potential contributions of the business sector, in general, and MNCs, in particular, to poverty reduction. Next, five distinct business strategies on poverty alleviation and their possible variations were presented, being Business-as-Usual, Corporate Philanthropy, Microfinance, Bottom of the Pyramid and Social Entrepreneurship. These strategies were subsequently examined and classified based on their poverty eradication potential. Afterwards, the topic on Cross-Sector Partnerships on poverty alleviation was introduced, being followed by a comparable analysis and classification. As a proceeding step, the simultaneous combination of the presented strategies as well as the execution of these strategies through a partnership approach was discussed and combination possibilities were classified. Ultimately, based on all prior classification of the presented approaches, namely Business-as-Usual, Corporate Philanthropy, Microfinance, Bottom of the Pyramid, Social Entrepreneurship and Cross-Sector Partnerships, as well as their variations and simultaneous combination possibilities, a firm's Poverty Alleviation Portfolio Approach was comprised. Building upon the prior sub-classifications, these Poverty Alleviation Portfolios were classified, according to a framework developed by van Tulder and van der Zwart (2006), as inactive, reactive, active or proactive approaches.

In turn, in order to investigate the actual contributions of MNCs toward poverty alleviation from a practical viewpoint, a content analysis of the websites and the CSR reports of the Fortune Global 100 firms, with regard to their reported poverty alleviation initiatives, was conducted. Hereby, engagement in the previously presented business approaches on poverty reduction, as well as their variations and simultaneous combination possibilities was examined. To facilitate the interpretation of the information, the retrieved company statements were codified into specified categories. As a proceeding step, the 100 MNCs were classified according to the developed classification framework from the theoretical part. More precisely, Poverty Alleviation Portfolio scores per firm were obtained based on the joint sub-classification scores for engagement in Corporate Philanthropy, Microfinance, Bottom of the Pyramid, Social Entrepreneurship, Cross-Sector Partnerships as well as the Combination of Business Strategies and the Combination of Business Strategies with Cross-Sector Partnerships. Consequently, the final score, being the sum out of these seven previous scores, determined a company's ultimate Poverty Alleviation Portfolio Approach, which was classified as inactive, reactive, active or proactive.

To summarize, the analysis conducted in the theoretical part of this paper has revealed that the poverty alleviation potential of companies, in general, and MNCs, in particular, is great, even though it is acknowledged that the private sector alone is insufficient to entirely eradicate poverty. Thus, additional, complementary initiatives by the public sector and civil society are required as well. However, whereas the claim *that* MNCs can substantially contribute to poverty reduction seems to be widely accepted, widespread ambiguity remains on *how* – and thus through which business initiatives or portfolio approaches – this contribution could best be realized.

This thesis therefore provided an overview and critical reflection of the main academic findings regarding the poverty-alleviating impact of distinct business approaches. The performed analysis reveals that the examined approaches differ in their effectiveness with regard to poverty reduction. Moreover, a simultaneous combination of approaches which focus on addressing poverty is possible and considered as beneficial for combating poverty. With respect to the resulting classification, an inactive Poverty Alleviation Portfolio Approach is characterized by sole engagement in Business-as-Usual and no undertaking of poverty-focused strategies or partnering activities. Furthermore, a reactive Poverty Alleviation Portfolio Approach relates to the engagement in classical Corporate Philanthropy and narrow Microfinance, a limited partnering approach and a restricted overall number of initiatives on poverty reduction, primarily pursued in isolation. Subsequently, the pursuit of an active Poverty Alleviation Portfolio Approach corresponds to the engagement in strategic Corporate Philanthropy, broad Microfinance, narrow Bottom of the Pyramid and narrow Social Entrepreneurship, a moderate partnering approach as well as a modest number of initiatives mostly pursued in combination. Ultimately, a proactive Poverty Alleviation Portfolio Approach resembles engagement in inclusive Bottom of the Pyramid and broad Social Entrepreneurship, an extensive partnering approach with a substantial number of undertaken initiatives, entirely pursued in partnership. Accordingly, it can be expected that the more proactive a firm's Poverty Alleviation Portfolio Approach, the more extensive its engagement in effective, poverty-related initiatives, and hence the larger its potential contribution to poverty reduction.

Considering the practical part of this thesis, the findings indicate that although the overwhelming majority of selected MNCs do engage in at least one initiative on poverty alleviation, these pursued approaches frequently classify at most as reactive. Hence, firms tend to favor the less effective, quick-fix options, such as classical Corporate Philanthropy, over more sustainable, proactive poverty-alleviation initiatives, such as inclusive Bottom of the Pyramid, broad Social Entrepreneurship or extensive partnering approaches. Additionally, even though examples for all possible, simultaneous combinations of approaches could be obtained, the number of combined initiatives on the whole is considerably low. Accordingly, the more proactive the approach, the lower the number of engaged companies, with 59 firms classifying as inactive, 32 as reactive, 8 as active and 1 sole company pursuing a proactive Poverty Alleviation Portfolio Approach. Further analyses of these results reveal that contextual company factors, such as sector, region of origin or company narrative seem to at least partially explain the divergence in the ultimate classifications of MNCs.

All in all, this thesis reveals that the potential contribution of MNCs to poverty alleviation, in particular through a proactive Poverty Alleviation Portfolio Approach, *could* be substantial in theory. However, research findings indicate that, in practice, the overwhelming majority of MNCs *do not* regard poverty as a priority and are primarily classified as pursuing an inactive or a reactive Poverty Alleviation Portfolio Approach. As a consequence, at present, MNCs on the whole do not explore their full poverty alleviation potential, thereby leaving substantial room for improvement, in order to ultimately maximize their positive contribution to the elimination of poverty.

4.4 Limitations and Suggestions for Further Research

After outlining the contributions of the above research, this section highlights limitations of the employed approach. Moreover, several suggestions for further research are presented to stimulate additional studies on the topic.

4.4.1 Limitations

To begin, it has to be mentioned that the topic of poverty alleviation is a particularly broad, complex and widely studied field of research. Poverty is addressed within numerous academic fields, including business, development, economics or political sciences. Therefore, in order to gain sufficient understanding of the topic and its diverging facets, an extensive dive into existing literature review is required. Furthermore, opinions and research findings frequently tend to be controversial which make an objective analysis of the topic challenging.

The in-depth literature review on business approaches to poverty alleviation revealed, that there prevails a substantial gap in a broad comparison of poverty-related initiatives in the area of business. This thesis tries to address that gap by classifying companies' Poverty Alleviation Portfolio Approaches. Thereby, a helicopter perspective was employed which puts different conceptions into comparison, in order to highlight similarities and differences and to present a 'big picture' with the topic placed in the wider context. However, this approach also has its disadvantages due to the fact that a substantial number of topics are covered and hence, in-depth elaboration of individual themes is relatively limited.

A further limitation is the fact that, in order to restrict the scope of the thesis, certain related perspectives and approaches were neglected. First of all, poverty alleviation was reviewed by focusing on a business perspective via the potential contribution of companies to development. Hence, the roles of other societal actors, such as government and NGOs have only been superficially touched upon. Furthermore, during the analysis of the strategies on poverty alleviation, emphasis was placed on the impact of these initiatives on the poor. Consequently, the positive and negative consequences of such undertakings of the business itself were largely ignored, partially also due to the fact that this has already been widely researched in past studies. The final major restriction in scope refers to the fact that only a selective number of business initiatives on addressing poverty were covered. Thus, additional initiatives, which turned out to be less relevant as a consequence of the structured literature review, were widely ignored.

A final essential limitation is the fact that this thesis attempted to classify and thereby compare several business initiatives on poverty reduction in theoretical and in practical terms. Such a classification is in itself a partially subjective undertaking. Even though special attention was paid to considering a wide variety of factors and research results in addition to clarifying the reasoning and steps being undertaken, full objectivity of the resulting classification cannot be guaranteed. Moreover, the descriptions of the companies' actual initiatives may also have been partially inaccurate since companies might not display all relevant information on the websites or CSR reports. Thus, certain undertakings which might not have been listed were consequently ignored which in turn could impact the accuracy of the ultimate classification.

4.4.2 Suggestions for Further Research

Based on the previously outlined limitations of this study, the following suggestions for further research can be made. First of all, a substantial research gap in terms of empirical findings on the implications on poverty of distinct business approaches, as well as their combinations and variations, was revealed. Consequently, more in-depth studies are required on the individual *and* joint impact of such initiatives on the poor, both in theory and in practice. Moreover, empirical research could be conducted, that combines the concept of Cross-Sector Partnerships with prevailing business strategies on poverty reduction and thereby investigates the effects of using these approaches in conjunction. Hereby, additional business strategies which have been excluded in this study could be incorporated as well. In this regard, particularly a research which employs companies' Poverty Alleviation *Portfolios* as unit of analysis would be value-adding.

Moreover, the primary research in this study solely analyzed and classified actual undertakings of MNCs based on data gathered from the companies' websites and sustainability reports. This might result in a non-all-encompassing description of actual initiatives. Further research could therefore enrich the findings of this study via additional data gathering methods, such as interviews or questionnaires, to increase the active participation of companies. Besides, the current focus, which has been placed on MNCs, could be broadened by studying SMEs as well. Moreover, future research could add to the findings of this study, by empirically exploring underlying reasons for variations in companies' Poverty Alleviation Portfolio Approaches. Thereby, studies could build upon the suggested causal explanation framework by empirically testing the impact of company sector, region of origin and company narrative on a company's classification. Ultimately, it should also be mentioned that the proposed Poverty Alleviation Portfolio classification is by no means regarded as finite, but as a starting point for further, joint analyses. Hence, additional research is encouraged to further revise, verify or enrich the employed classification framework of this thesis.

5 Annex

Appendix A: Theoretical Part – Structured Literature Search Overview

Keywords	AND Poverty		AND Poverty AND Partnership		AND Poverty		AND Poverty AND Partnership		AND Poverty		AND Poverty AND Partnership		AND Poverty		AND Poverty AND Partnership	
Database	Ebsco				Abi Inform/ Proquest				Science Direct				Scopus			
Search criteria	all text	abstract	all text	abstract	all text	abstract	all text	abstract	all text	abstract	all text	abstract	all text	abstract	all text	abstract
Partnership																
Partnership	7892	106	x	x	7182	107	x	x	2769	34	x	x	4848	359	x	x
Collaboration	4945	21	x	x	5085	28	x	x	2385	19	x	x	2051	162	x	x
"Cross-sector collaboration"	21	1	x	x	20	1	x	x	7	0	x	x	11	1	x	x
"Cross-sector partnership"	5	1	x	x	14	1	x	x	9	0	x	x	29	3	x	x
"Cross-sectoral partnership"	4	0	x	x	2	0	x	x	4	0	x	x	5	0	x	x
"Multi-sector partnership"	2	0	x	x	2	0	x	x	1	0	x	x	2	0	x	x
"Tripartite partnership"	7	0	x	x	8	0	x	x	5	0	x	x	2	0	x	x
"Public-private partnership"	372	4	x	x	325	5	x	x	349	5	x	x	445	34	x	x
"Co-creation"	72	0	x	x	69	0	x	x	14	0	x	x	21	1	x	x
"Partnership for development"	153	3	x	x	0	0	x	x	0	0	x	x	0	0	x	x
"Partnership portfolio"	1	0	x	x	0	0	x	x	1	0	x	x	0	0	x	x
Sum selected partnership articles	14	136			12	142			27	58			9	560		

Table 5-1: Literature Search Results per Database and Keyword for Partnerships (Pre-Selected Articles are marked in Yellow and Orange)

Keywords	AND Poverty		AND Poverty AND Partnership		AND Poverty		AND Poverty AND Partnership		AND Poverty		AND Poverty AND Partnership		AND Poverty		AND Poverty AND Partnership	
Database	Ebsco				Abi Inform/Proquest				Science Direct				Scopus			
Search criteria	all text	abstract	all text	abstract	all text	abstract	all text	abstract	all text	abstract	all text	abstract	all text	abstract	all text	abstract
Business strategies																
"Inclusive business"	9	0	7	0	5	0	6	0	0	0	0	0	8	2	3	1
"Bottom-of-the-Pyramid"	272	20	133	2	157	10	86	2	40	2	25	0	340	19	76	2
"Base-of-the-Pyramid"	80	8	41	0	69	4	41	0	21	4	11	0	103	14	27	0
"Social business"	66	6	32	0	31	1	14	0	13	1	11	0	53	3	6	0
"Social enterprise"	151	2	84	0	119	1	72	0	56	1	36	0	131	15	42	2
"Social venture"	34	0	22	0	22	0	17	0	13	0	10	0	27	1	10	0
"Social venturing"	0	0	0	0	1	0	1	0	1	0	1	0	1	0	1	0
"Sustainable livelihood"	101	1	47	0	73	2	31	0	228	18	85	0	756	83	127	6
Philanthropy	1090	8	608	1	505	3	273	1	174	1	75	0	501	25	98	1
"Sustainable global enterprise"	6	0	3	0	6	0	6	0	0	0	0	0	3	0	0	0
"Subsistence market"	5	2	2	1	3	1	2	0	12	3	5	1	53	4	3	1
"Subsistence marketplace"	2	1	0	0	2	0	0	0	14	5	5	1	27	0	7	0
Microfinance	792	67	262	0	367	38	103	1	243	25	82	0	1115	226	96	5
Microcredit	485	26	147	2	331	13	141	0	157	8	57	0	585	83	37	1
"Inclusive banking"	2	0	1	0	1	0	0	0	0	0	0	0	0	0	0	0
"Fair trade"	669	13	298	0	340	7	149	0	159	1	62	0	400	30	95	2
"Code of conduct"	458	1	249	0	258	0	119	0	256	1	99	0	285	3	73	0
Label	3543	7	817	0	1963	2	456	0	1580	3	327	0	497	30	57	0
"Affordable products"	0	0	0	0	13	0	14	0	5	0	4	0	1	0	0	0
"Child labor"	719	29	179	0	547	12	17	0	421	24	94	0	599	97	30	0
"Decent wages"	0	0	0	0	25	0	19	0	15	0	0	0	4	1	0	0
"Working poor"	922	25	227	0	468	7	104	0	164	3	25	0	675	86	41	1
"Ethical business practice"	0	0	0	0	7	0	7	0	14	0	7	0	4	0	0	0
"Foreign Direct Investment"	2672	28	922	2	895	6	244	0	604	6	161	0	962	48	94	4
"Job creation"	1483	19	530	0	562	12	196	0	369	6	129	0	347	48	42	0
Spillovers	1957	19	338	0	395	10	152	0	623	8	126	0	716	34	50	1
"Technology transfer"	1160	8	554	0	498	5	222	1	359	7	133	0	402	27	65	1
"Knowledge transfer"	358	0	225	0	182	0	91	0	153	0	70	0	121	4	21	2
"Corporate Social Responsibility"	869	19	471	0	666	8	337	0	212	11	112	2	534	38	173	4
"Pro-poor strategy"	11	2	3	1	3	1	1	0	11	1	4	0	15	5	7	1
Sum selected business strategy articles	19	311	16	9	25	143	23	5	6	139	36	4	21	926	27	35

Table 5-2: Literature Search Results per Database and Keyword for Business Strategies (Pre-Selected Articles are marked in Yellow and Orange)

Selected Articles	Sum all selected business/partnership AND poverty	Sum abstract business/partnership AND poverty	Sum all selected business AND Poverty AND partnership	Sum abstract business AND Poverty AND partnership
Business strategies				
"Inclusive business"	24	2	17	1
"Bottom-of-the-Pyramid"	51	51	6	6
"Base-of-the-Pyramid"	30	30	0	0
"Social business"	11	11	6	0
"Social enterprise"	19	19	2	2
"Social venture"	1	1	10	0
"Social venturing"	3	0	3	0
"Sustainable livelihood"	104	104	6	6
Philanthropy	37	37	3	3
"Sustainable global enterprise"	15	0	9	0
"Subsistence market"	10	10	15	3
"Subsistence marketplace"	10	6	13	1
Microfinance	356	356	6	6
Microcredit	130	130	3	3
"Inclusive banking"	3	0	1	0
"Fair trade"	51	51	2	2
"Code of conduct"	5	5	0	0
Label	42	42	0	0
"Affordable products"	6	0	4	0
"Child labor"	162	162	0	0
"Decent wages"	5	1	0	0
"Working poor"	121	121	1	1
"Ethical business practice"	11	0	14	0
"Foreign Direct Investment"	88	88	6	6
"Job creation"	85	85	0	0
Spillovers	71	71	1	1
"Technology transfer"	47	47	2	2
"Knowledge transfer"	4	4	2	2
"Corporate Social Responsibility"	76	76	6	6
"Pro-poor strategy"	12	9	17	2
Sum selected business strategy articles	**1590**	**1519**	**155**	**53**
Partnership				
Partnership	606	606	x	x
Collaboration	230	230	x	x
"Cross-sector collaboration"	10	3	x	x
"Cross-sector partnership"	14	5	x	x
"Cross-sectoral partnership"	15	0	x	x
"Multi-sector partnership"	7	0	x	x
"Tripartite partnership"	22	0	x	x
"Public-private partnership"	48	48	x	x
"Co-creation"	1	1	x	x
"Partnership for development"	3	3	x	x
"Partnership portfolio"	2	0	x	x
Sum selected partnership articles	**958**	**896**		

Table 5-3: Literature Search Results per Keyword (Pre-Selected Articles are marked in Orange)

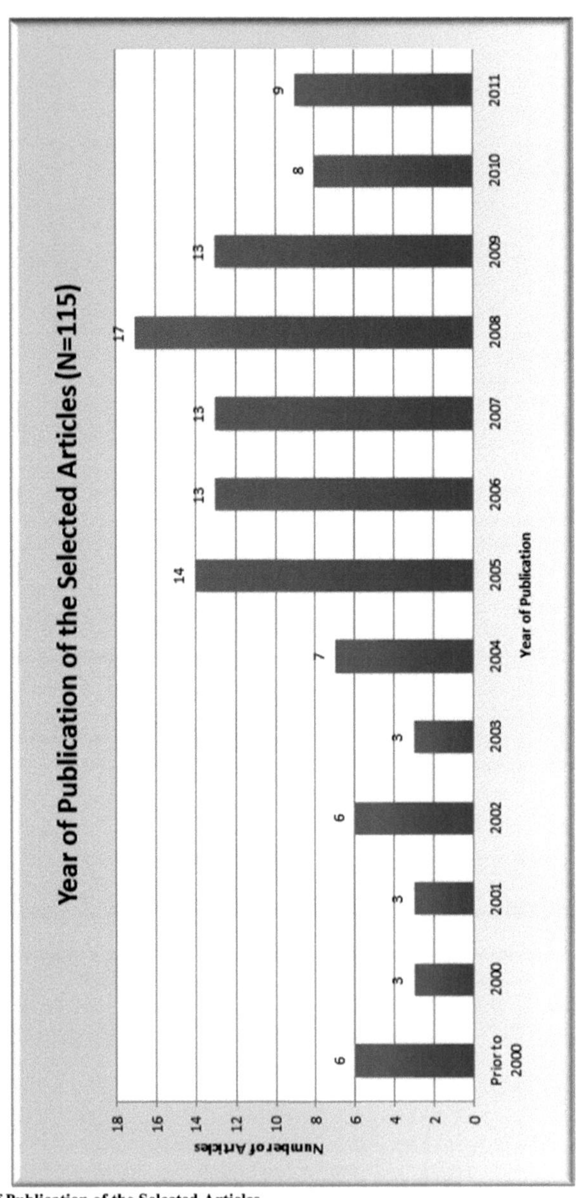

Figure 5-1: Year of Publication of the Selected Articles

Appendix B: Theoretical Part – Literature Review Citations Section 2.2

Reference	Citations on the Importance of the Business Sector for Poverty Alleviation
Barney, 2003	"The private sector is increasingly seen to have a role to play in helping to meet international development targets on poverty reduction." (p. 255). "Corporate responses to a greater or lesser extent were encouraged to integrate social issues into core rather than peripheral business practice" (p. 256). "Adopting a more positive approach to development issues demands a shift in focus from peripheral to core business practice. It also demands a willingness to seek and address causes and not symptoms." (p. 262-263).
Farias and Farias, 2010	"Traditional approaches to poverty alleviation have essentially focused on the role of government and NGOs in providing infrastructure, welfare programs and programs to generate employment. However, more recent approaches have begun to emphasize the role of business in achieving sustainable development." (p.249). "Enterprise-based solutions to poverty will help speed up economic development for the poor." (p. 255).
Frynas, 2008	"Both development agencies and academic scholars have in recent years made claims about the positive role that CSR could play in contributing to international development goals, such as poverty alleviation [...] The linking of CSR to international development is a hugely significant development, in that firms are not simply expected to act responsibly, but to play an important role in public interventions, such as the UN's MDGs." (p.274). "The expectations placed on the private sector have undoubtedly grown." (p. 275).
Humphrey, 2006	"Achieving [poverty alleviation and other development goals] involves interventions at multiple levels. The private sector has a role to play." (p. 46).
Item, 2007	"It is evident that an expanding range of stakeholders is defining more broadly the roles and responsibilities of business in society, in particular global business [...] Companies are increasingly being obliged by a growing range of stakeholders to play a positive role in society as corporate citizens. The result is that globalization is redrawing the boundaries of responsibility for business and in some areas of social development [...] merging corporate responsibilities with those of government, requiring business to address social development goals, increasingly in partnership with other societal actors [...] Sustainable development requires business organizations to take their economic, social and environmental responsibility seriously." (p. 11-12).
McKague et al., 2004	"Sustainable and equitable growth is central to poverty reduction and this economic growth is best achieved through the private sector. Never before has the consensus been greater about the potential private sector contribution to poverty alleviation and sustainable development." (p. 1). "Market forces, not government, are needed for the private sector to play its poverty-alleviating role." (p. 106).
Sasse and Trahan, 2006	"Corporations are becoming more and more inclined to engage in the social sector, whether it is as stated as fulfilling their social responsibility, promoting philanthropy, or acting as a responsible citizen." (p. 30).

Reference	Citations on the Importance of the Business Sector for Poverty Alleviation
Kelley, Werhane and Hartman, 2008	One innovative poverty-reduction initiative: "The transfer of one segment of the responsibility for global poverty from traditional international development practices to pioneering, private, for-profit organizations. This transfer can produce appropriate and effective incentives, stakeholder interest maximization, economic growth, and the potential for the reduction of both poverty and the unfulfilled needs of the abject poor [...] Alleviating poverty is a growth opportunity for the poor and for business"(p. 6). "The MDG to reduce poverty can be achieved only with substantive contributions of businesses as well as government and international organizations in ways that are sustainable and poverty alleviation is a profitable endeavor both for companies and for the poor when focused on corporate core competencies linking new market development with job creation and new product branding, not on the fringe of CSR in the form of charitable contributions." (p. 28).
Kolk, van Tulder and Kostwinder, 2008	"In the international discussion on furthering development, the potential contribution of companies is frequently mentioned. The 2002 Monterrey consensus stipulated the role of business in achieving development goals." (p. 263).
Newell and Frynas, 2007	"Business plays an increasingly important role in development. This is linked both to the decline in confidence in the role of the state as an agent for development and to global deregulation from the 1980s resulting in a more limited role of the state in the economy. Key development functions traditionally associated with the state are now performed by a wide range of Civil Society and market actors. As providers of goods and services, as employers, as investors and increasingly as shapers of developing countries' policies, there is no doubt that the private sector is central to efforts to tackle poverty." (p. 672).
Prieto-Carrón et al., 2006	"The increases in FDI in some developing countries, the promotion of private sector development and the rise of the CSR discourse [...] have played a central role in shifting the terms of the debate on the relationship between business and poverty. Formerly, the issue used to be how business causes poverty; now business is increasingly portrayed as part of the solution to the problem of world poverty through the promotion of free markets and the incorporation of SMEs in global supply chains, with help from large companies." (p. 980).
Raufflet, Berranger and Aguilar-Platas, 2008	"Over the last few years there has been an increasing interest in order to make business part of the solution to poverty. Several approaches have been developed, proposed, and diffused from both practice and research. Business does not operate in a vacuum, nor can it ignore local milieus it may contribute to positively – in addressing poverty-related issues for instance." (p. 56).
Raufflet, Berranger and Gouin, 2008	"Over the last decade, management researchers and business managers and leaders have displayed an unprecedented interest in poverty-related issues from several perspectives [...] These approaches share similar concerns: they aim to reconcile market and business approaches with pressing poverty reduction and development challenges, more particularly in the developing world." (p. 546).
McMullen, 2011	"Business has the potential to do far more to solve the issues of poverty than any number of government aid programs." (p. 206).

Reference	Citations on the Importance of the Business Sector for Poverty Alleviation
Sharp, 2006	"Since the late 1990s there has been a concerned move by international and national development agencies to add development to the list of business corporations' social responsibility." (p. 216).
Singer, 2006	"It is often said that the best cure for poverty in any region of the world lies in stimulating more business activity and start-up ventures. Although wealth creation does not necessarily imply poverty reduction, it is obvious that business activity alleviates some instances of income poverty." (p. 225-226).
Stefanovic, 2007	"The leading companies of 2020 will be those that provide goods and services and reach new customers in ways that address the world's major challenges – including poverty." (p. 6).
UNDP, 2004	"The private sector, particularly large local companies and MNCs, should contribute to accelerated economic development and to poverty alleviation." (p. 41).
van Tulder and van der Zwart, 2006	"Firms are not necessarily held responsible for (global) societal problems, but they are increasingly considered to be part of the solution" (p. 19).
Wadham, 2009	"Business is part of the problem (of poverty) but [it is now also] widely accepted that business is part of the solution." (p. 3).
Walsh, 2005	"People are looking for fresh solutions to the deep and long-standing problem of poverty. Many eyes turn to the corporation for help. Corporations are the most powerful institutions on the transnational stage today. Their wealth and incredible reach into factor and product markets worldwide mark them as ready targets for appeal [...] An enduring and contentious question, however, is how we might deploy these capabilities to address the problem of human misery." (p.473).
Warhust, 2005	"The roles and responsibilities of business in society, in particular global business, are being defined more broadly by an expanding range of stakeholders" (p. 151). "Society is asking business to contribute more directly to poverty alleviation and other development goals." (p.152). "Stakeholders are increasingly requiring businesses, especially MNCs, to be a positive force, to contribute to broader societal development goals and to work in partnership with others to solve humanitarian crises and endemic problems facing the world, such as disease and poverty." (p.153). "Business and its activities are inextricably both, part of the problem as well as the solution." (p.155).
WBCSD, 2006	"Business is now accepted as a key partner and solution-provider [...] and it itself recognizing the needs and opportunities presented by the world's poor people." (p. 2).

Reference	Citations on the Importance of the Business Sector for Poverty Alleviation
World Business Council for Sustainable Development, 2010	"Business is good for development and development is good for business." (World Bank, 2010, p. 10). "History shows that business not government develops a nation economically. Governments create the frameworks that encourage – or hinder – that development; but it is the private sector that generates entrepreneurship, creates employment and builds wealth. Companies, moving beyond conventional wisdom and working with new partners, have an unprecedented opportunity to help people to lift themselves out of poverty and into market economies. [...] Business, working in a spirit of 'enlightened self-interest' can improve the development path of billions of people, by facilitating their access into the marketplace, by finding new ways to address the needs of the poor and helping them into mainstream economic activity. A growing body of evidence indicates that intelligent engagement will also result in new revenues and profit." (p. 10). "There are many positive ways for business to make a difference in the lives of the poor – not through philanthropy, though that is also very important, but through initiatives, that, over time, will help to build new markets." (Kofi Anan, 2010, p. 11).

Table 5-4: Citations on the Importance of the Business Sector for Poverty Alleviation

Reference	Citations on the Importance of MNCs for Poverty Alleviation
Heslam, 2007	"Commercial enterprise faces unprecedented opportunities to be an agent of positive social [...] transformation in the contemporary world. It is [...] because global business enterprise demonstrates an ability to lift people out of poverty [...] Ignoring the role business can play in reducing poverty is a major oversight." (p. 132-133) "MNCs are the key agents of globalization, responsible for the vast majority of international trade. Their operations and investments in developing countries have helped millions of people to escape from poverty and their actions can greatly influence SMEs and all other parts of host countries." (p.135).
Kolk and van Tulder, 2006	"In the debate on how to combat poverty, the positive role of MNCs is frequently mentioned nowadays, although doubts and criticism remain. [...] This paper analyzes MNCs' policies on their poverty-alleviating potential. 'Frontrunner' MNCs turn out to be not very outspoken, especially not on those issues that have the largest potential to help alleviate poverty." (p. 789).
Oetzel and Doh, 2009	"On the one hand, some researchers are optimistic about the impact of MNEs on developing countries. MNEs are important agents, they argue, for promoting economic growth since they complement domestic savings, transfer technology and management skills, increase competition, and stimulate entrepreneurship. On the other hand, another group of researchers opposed this view suggesting that MNEs are more likely to crowd out local firms, use technology that is inappropriate for local circumstances, actively constrain potential technology spillovers and reduce the domestic capital stock and tax basis due to transfer price manipulation and excessive profit repatriation." (p. 108).
Pless and Maak, 2009	"There is widespread agreement in both business and society that MNCs have an enormous potential for contributing to the betterment of the world. In fact, a discussion has evolved around the role of business 'as an agent of world benefit'. At the same time there is also growing willingness among business leaders to spend time, expertise and resources to help solve some of the most pressing problems in the world, such as poverty." (p. 59).

Reference	Citations on the Importance of MNCs for Poverty Alleviation
Stoner and Wankel, 2007	"The possibilities for creating and taking innovative approaches to reduce poverty are seemingly endless [...] In terms of 'sizes', one set of innovations focuses on MNCs and addresses the ways they can contribute to reducing poverty." (p. 9) "Companies is the for-profit private sector – seeking to achieve traditional economic success – are also seen as major contributions to the reduction of poverty [...] the role commercial businesses can, should, and perhaps even must play in reducing poverty [...] such business should therefore be considered part of the solution to poverty, rather than merely part of the problem." (p. 12-13).
van Tulder and da Rosa , 2010	"MNEs not only can bring development through their direct investments, but also through the indirect effect of their investments through their linkages with local firms." (p. 2). "New business models have become available that approach the issue of poverty from either a positive or a negative side." (p. 5).
Yunus, Moingeon and Lehmann-Ortega, 2010	"Established MNCs have recently shown some interest in the [...] fight against poverty as part of a more general emphasis on CSR. However, shareholder value maximization remains the rule in the capitalist system and – clearly – the reconciliation of this with social objectives is often problematic. [...] However, research has shown that, if managed strategically, CSR projects can indeed pay off, both socially and financially." (p. 309).

Table 5-5: Citations on the Importance of the MNCs for Poverty Alleviation

Reference	Citations on the Potential Contribution of the Business Sector for Poverty Alleviation
Goldsmith, 2011	„A growing movement commends a targeted business-oriented approach as the best chance to reduce global poverty. It offers the promise of dual returns – empowerment of the world's neediest citizens and earned income for investors in pro-poor ventures at the same time [...] While they differ in detail, these business-oriented, anti-poverty strategies are held to be potentially more stable financially than private charity or public assistance, and less likely to create dependency among low-income beneficiaries." (p. 15).
Heslam, 2007	"The resulting irony is that the positive impact business could have on poverty through its core activities, which is generally much greater than through its CSR initiatives, is overlooked." (p. 134).
Lyon and Bertotti, 2007	"Provision of jobs is the most direct effect of business growth [...] The greatest impacts on poverty reduction occur when businesses can develop new innovative markets, employing the less skilled people and creating escalator jobs that increase skill levels [...] This study has also shown that poverty-reduction innovation in enterprises does not relate only to providing jobs, products and services. Firms may also be innovative in terms of HR development, increasing job quality, training and community involvement, all of which can contribute to poverty reduction." (p. 196).
McFalls, 2007	"Increasingly, major corporations are being targeted to play an active leadership role in promoting sustainable development [...] Corporations can and should find win-win business opportunities in developing countries that address specific development goals as well as satisfy the commercial interest of the firm." (p. 2). "Entrepreneurship on a massive scale that will uplift the poor and yield profits to the firms that engage in what becomes a win-win for business and society." (p. 18).

Reference	Citations on the Potential Contribution of the Business Sector for Poverty Alleviation
UNDP, 2004	"The private sector can alleviate poverty by contributing to economic growth, job creation and poor people's incomes. It can also empower poor people by providing a broad range of products and services at lower prices." (p. 1). "The private sector, particularly large local companies and MNCs must realize that it can contribute to accelerated economic development and to poverty alleviation [...] through its knowledge, expertise, resources and relationships." (p. 3). "The private sector is central to the lives of the poor and has the power to make those lives better. It is about using the managerial, organizational and technological innovation [...] to improve the lives of the poor." (p. 5).
Pfeffermann, 2001	"Little has been published about the development impact of private enterprises per se." (p. 42). "Job creation is a major – probably the major – path out of poverty. The most sustainable job creation is by firms. [...] In almost all developing countries [...] private enterprises are the main source of new jobs." (p. 43). "Private firms also contribute to development in other ways that are crucial to economic development and poverty reduction (i.e. via tax revenue generation, competitiveness and affordable, high quality products)." (p. 44).
WBCSD, 2004	"History shows, that business – not government – develops a nation economically. Governments create the frameworks that encourage – or hinder – that development; but it is the private sector that generates employment. [...] All companies, regardless of their industry, can help stimulate local markets and enable the poor to become active participants in these markets, as customers and entrepreneurs. Designing clever business models to address this challenge will also open new avenues of growth for the company." (p. 6).

Table 5-6: Citations on the Potential Contributions of the Business Sector for Poverty Alleviation

Reference	Citations on the Potential Limitations of the Business Sector for Poverty Alleviation
Campbell, 2005	"The economic impact of corporate activities and the relationship of these impacts to broader socio-economic development goals are little understood. [...] While deliberate attempts by companies to reach underserved markets have been profitable and therefore sustainable, for some companies, these have tended to be marginal to core business activities. [...] Unless companies manage and maximize the economic impact of their core business [...] improvements in these issues are likely to remain at the margins." (p. 413). "Companies should be accountable for any impacts on society that are directly and uniquely attributed to their business activities." (p. 419)
Frynas, 2008	"The article suggests that many recent claims about the positive contribution of CSR to international development are unjustified based on four arguments: (1) Lack of empirical evidence, (2) Analytical limitations of CSR, (3) The constraints of the business case for CSR and (4) Unresolved governance questions." (p. 274). "As a consequence of liberalization and deregulation, firms are now being called upon to go beyond their traditional role of generating economic growth toward playing a more direct role in alleviating poverty and other development goals [...] However, while private sector development initiatives can perhaps be beneficial for specific firms in terms of reputation effects or new product development, we know relatively little about their developmental benefits." (p. 275). "It has been shown that even MNCs regarded as CSR best performers have very limited potential to contribute

Reference	Citations on the Potential Limitations of the Business Sector for Poverty Alleviation
	positively to poverty alleviation.“ (p. 278).
Hermann, 2003	“There remain questions about how effectively the private sector can contribute to addressing the many social and development challenges facing the world's poorest nations. In addition, little is known about how to measure private sector efforts to address poverty as the motivations for engaging in poverty reduction differ significantly.“ (p. 1).
Kelley, Werhane and Hartman, 2008	“The socially conscious are often skeptical of business interests (in poverty alleviation) considering well-publicized patterns of exploitation..“ (p. 11).
Kolk and van Tulder, 2006	“While the potential and opportunities of MNCs in relation to poverty alleviation are receiving considerable attention, the exact components for fulfilling such a role, in the context of international effort to delineate poverty and the contribution of MNCs in this regard, have hardly been addressed [...] It has been difficult to assess the direct and indirect effects of MNCs' activities on developing countries overall.“ (p. 790).
Lyon and Bertotti, 2007	“At present there is a lack of robust empirical evidence to demonstrate the actual contribution of enterprises to the reduction of poverty and deprivation.“ (p. 176)
McMullen, 2011	“For market-based approaches to produce development outcome there must be some degree of charity to supplement the profit motive.“ (p. 198).
Merino and Valor, 2011	“The claims that business and CSR strategies can be effective in fighting poverty face major challenges, in particular the wide divergence of understandings about the notion and implementation of CSR, and the absence of clear understandings of underlying ideological bases concerning business and poverty.“ (p. 157). “Today it is accepted that although FDI may have a positive influence on national economic growth, it may not have a positive influence on social and environmental indicators. Companies are still regarded as the driving force of economics but we have gone beyond the idea that corporate presence as such is an effective aid to the realization of the MDGs. [...] Corporate presence per se is not sufficient to eradicate poverty. [...] Strengthening the private sector is a necessary condition: The private sector mitigates poverty by means of its contribution to economic growth. [...] However, it is understood that corporate activity has negative consequences, which unless minimized will not make an effective contribution to the reduction of poverty. [...] CSR is by nature development carried out by the private sector and results in being the perfect complement of efforts made by the government and multilateral institutions” (p. 159). “Critics contend that there is no such clear relationship between CSR and poverty reduction.“ (p. 165).
Newell and Frynas, 2007	“Poverty is much more than lack of income. Contributions to poverty alleviation which rest solely of the potential of business to promote growth or provide jobs are therefore limited in addressing the underlying causes of poverty which exclude people from the labor market in the first place. [...] Different models of CSR have impacts on different types of poverty. [...] Establishing which models benefit poorer groups most is the challenge.“ (p. 673) “Our previous research suggests that CSR initiatives work for some firms, in some places, in tackling some issues, some of the time. The challenge is to explore the potential and limitations of CSR in specific settings.“ (p. 674) “A holistic view of poverty is required in which CSR initiatives have a contribution to make, but governments and donors

Reference	Citations on the Potential Limitations of the Business Sector for Poverty Alleviation
	have to be realistic about what can be achieved by them in isolation from their own efforts to tackle poverty.“ (p. 676). ”It is clearly unrealistic to expect businesses in a globalized capitalist economy to operate as if poverty alleviation were their main objective. The greatest contribution CSR activities can make is through reinforcing state-led development policy. [...] Encouraging firms to take seriously their social and environmental responsibility is to be encouraged. The business of business is no longer just business if indeed it ever was. Expecting too much of CSR, particular regarding its contribution to tackling poverty, however, is unrealistic.“ (p. 678-679).
Prieto-Carrón et al., 2006	“Numerous arguments have been made about the potential benefits accruing to companies, workers and community members that engage in CSR activities, but there is clearly a need to generate more comparative evidence on the social and environmental impacts of such initiatives.“ (p. 981).
Sharp, 2006	“Some analysts suggest that corporate-driven social responsibility initiatives offer a new and potentially bright prospect of addressing global poverty and underdevelopment effectively. There is a growing academic literature that examines this proposition carefully.“ (p. 213). “There are inherent features which limit CSR’s ability to address poverty.“ (p. 214). “Most CSR initiatives are not intended to tackle questions of poverty and social exclusion. They aim at less ambitious goals of performance enhancement and image management.“ (p. 215). ”But the new business discourse of development sets out to limit what can be expected of corporations. [...] Those people who possess something, or some characteristic, that business wants are entitles to development in exchange [...] Those who are not members of host communities or stakeholders have no claim to recompense in any form.“ (p. 217).

Table 5-7: Citations on the Potential Limitations of the Business Sector for Poverty Alleviation

Appendix C: Theoretical Part – Literature Review Citations Section 2.3

Reference	Citations on Business-as-Usual
Citations on its Importance for Poverty Alleviations	
Ayanwu, 2006	"Direct investment is essential [...] in order to (a) enhance economic growth, (b) increase employment (c) reduce poverty." (p. 45).
Citations on its Potential Contribution to Poverty Alleviations	
Newell and Frynas, 2007	"Business-as-usual private investments can increase employment among the poor, provide new market opportunities for smallholders, increase the access of the poor to essential services or contribute to government taxes, which can in turn be spend on anti-poverty measures. Both the CSR and the non-CSR activities of the firm have a bearing on poverty even if the latter are far more significant in overall social and economic terms." (p. 674).
Citations on its Potential Limitations for Poverty Alleviations	
Gore, 2010	"Business-as-Usual is not a viable option now. What is required over the coming years is not tinkering to increase the effectiveness of poverty reduction efforts. Rather, there is a need for a new international development consensus and a new approach to international development cooperation." (p. 73).
Karher, Iyani and Shannon, 2007	"MNCs have important benefits for developing countries, including the capacity to build investment, the propensity for creating a sound economic structure and the promise of subsequently reducing poverty levels. [...] MNCs can do that profitably and sustainably" (Lodge and Wilson, 2006, p. 76). "Despite Lodge and Wilson's optimism, the track record is bleak insofar as developing a positive economic change in developing nations." (p. 76).
Devarajan, 2002	"Economic growth is not enough to combat poverty." (p. 2).
Norton, 2002	"It seems obvious that economic growth should reduce poverty, yet the issue remains controversial. Some scholars assert that economic growth does not eliminate poverty and may exuberate the problems of the poor." (p. 263).
Merino and Valor, 2011	"Corporate presence per se is not sufficient to eradicate poverty. [...] Strengthening the private sector is a necessary condition [...] however, it is understood that corporate activity has negative consequences, which – unless minimized – will not make an effective contribution to the reduction of poverty." (p. 159).
Jenkins, 2005	"It is surprising that research on the impact of FDI on poverty is so limited" (p. 532).

Table 5-8: Citations on Business-as-Usual and Poverty Alleviation

Reference	Citations on Corporate Philanthropy
Citations on its Potential Contribution to Poverty Alleviations	
Montgomery, Palma and Hoagland-Grey, 2008	"Private sector [...] companies operating in developing countries can positively affect the social, environmental and economic fabric of local communities by providing infrastructure services and through community investment programs focused on economic development, poverty reduction, social inequality reduction, and environmental improvements. Community investment programs, as part of the CSR concept, can take form of philanthropy or charitable donations, employee volunteerism and *partnerships* [...] By going beyond the minimal requirements to mitigate project-related negative social and environmental impacts, private sector companies can be models of social responsibility and good corporate citizenship." (p. 241).
Citations on its Potential Limitations for Poverty Alleviations	
Sasse and Trahan, 2006	"Corporate philanthropy is often well-meaning in providing short-term funding for a variety of social causes. [...] The efficiencies and competence of business to attack social causes are often overestimated; likewise, the actual unintended harms to society by way of corporate philanthropy are often underestimated for large corporations with deep pockets." (p. 37).
Thurman, 2006	"Traditional philanthropy displayed how stagnant and ineffective it really is [in reducing chronic global poverty]. The problem with philanthropy today is that too much attention is focused on counting receipts and too little on outcomes." (p. 18).

Table 5-9: Citations on Corporate Philanthropy and Poverty Alleviation

Reference	Citations on Microfinance
Citations on its Importance for Poverty Alleviations	
Hossain and Knight, 2008	"The underlying premise [of microfinance] is that charity, rather than eliminating poverty, will serve to compound it by taking away the individual's independence and will to overcome poverty. The Grameen bank proposes that the true answer to poverty is the realization of the creative potential of poor individuals." (p. 159).
Citations on its Potential Contribution to Poverty Alleviations	
Karher, Iyani and Shannon, 2007	"Regardless of the criticism, economic data points to Grameen's modest effects on lowering aggregate poverty rates." (p. 77).
Yunus, Moingeon and Lehmann-Ortega, 2010	"68 per cent of the families of Grameen Bank barrowers have crossed the poverty line." (p. 309).
Jones Christensen, 2008	"Results indicate that microcredit and the related microfinance movement are proven effective in poverty alleviation. Some statistics show that recipients of microloans lift themselves and their families above the poverty line within 12-14 loan cycles and certainly within one generation." (p. 151).
Hossain and Knight, 2008	"Our study revealed that microfinance continues to have a significant positive impact on poverty levels in Bangladesh, and is crucial to the socio-economic development of the poor [...] Micro-finance is firmly associated with the MDGs and poverty eradication [...] Despite its various limitations, the strength and success of microfinance cannot be ignored [...] Although poverty cannot be single-handedly alleviated by microfinance, it does play a large role in poverty reduction [...] The research found the Grameen microfinance model to be an effective model of financing for the poor and disadvantaged. This study did not observe any particular challenges or adverse findings with regard to high interest rates, exploitation of women, loan repayment, unchanging poverty levels and failure to cater

Reference	Citations on Microfinance
	effectively to the target groups. Rather, our observations suggest that the Grameen Bank is silently making a positive difference to the poor and their livelihoods." (p. 167-168).
Yunus and Jolin, 1999	"Microcredit is not a miracle that can eliminate poverty in one fell swoop. But it can end poverty for many and reduce severity for others. Combined with other innovative programmes that unleash people's potential, microcredit is an essential and in our search for a poverty-free world." (p. 172).
Hossain, 2000	"Microcredit has been identified as a key mechanism to fight poverty in societies." (p. 191).
Karlan and Valdivia, 2011	"A growing number of microfinance organizations are attempting to build the human capital of micro-entrepreneurs in order to improve the livelihood of their clients and help further their mission of poverty alleviation. [...] We find little or no evidence of changes in key outcomes such as business revenue, profits or employment; we nevertheless observed business knowledge improvements and increased client retention rates for the microfinance institution." (p. 510).
Citations on its Potential Limitations for Poverty Alleviations	
Pitta, Guesalaga and Marshall, 2008	"There is evidence that microloans have succeeded in aiming at the *BOP*. There is also evidence that many of the would-be entrepreneurs failed to capitalize on such credit. They got deeper into debt. The entrepreneurial skill that can lead to success is rare." (p. 398).
Eder and Öz, 2008	"The empirical evidence regarding the microcredit system is rather mixed, as it has been found out that microcredit dies not always result in the alleviation of income poverty." (p. 124).
Makita, 2006	"The existing literature tells us that the provision of *credit* and other services for self-employment was effective in reducing poverty, but was insufficient in helping the poor rise above poverty level. On the other hand, the advent of microcredit has increased differentiation among the poor through the exclusion of the poorest from credit provision and the increase of indebtedness [...] even if the poor can join they cannot necessarily use the credit effectively for their non-farm enterprise. [...] Some of the poor people have improved their livelihoods through credit programs, but some of them have had their livelihood worsened." (p. 219).
Johnson, 2009	"Findings from research argue that the impacts for poor people of expanding access arise predominately from the effects this has on economic growth rather than through giving them direct access to financial services." (p. 294). "Evidence from a number of studies suggests that the effects of direct access to financial services for the poor are less substantial than indirect effects via economic growth" (p. 295). "The evidence for the impact of microfinance on poverty reduction has been mixed in its findings. [...] While numerous studies have demonstrated positive impacts the fact that not all those who take on debt can raise their incomes, develop their asset base or reduce their vulnerability all of the time seems a result to be expected. [...] With this lack of overwhelming evidence, the evidence has been put on the methodological pitfalls of many past studies...and has resulted in the rising tide of randomized control approaches to finally prove the case. However, findings from recent randomized studies appear to support a view of the complexity and the heterogeneity of impact pathways and outcomes" (p. 296).

Reference	Citations on Microfinance
Shetty, 2010	"In recent years microcredit has grown rapidly based on its promise to alleviate poverty and empower women. However, as the microcredit industry has grown the initial emphasis on poverty alleviation and empowerment has changed; some critics argue that the industry now looks like commercial finance (requiring consistent profitable returns). [...] The findings suggest, that in general, the loans do not permanently move participants out of poverty; however, they do reduce some of the vulnerabilities associated with poverty." (p. 356). "It does not appear that microcredit can, by itself, fulfill its promise to empower women and alleviate their poverty." (p. 389-390).
Karnani, 2007a	"Most studies suggest that microcredit is beneficial but only to a limited extent. The problem lies not with microcredit but rather with microenterprises. With low skills, little capital and no scale economics, these businesses have low productivity and lead to meager earnings that cannot lift their owners out of poverty. Creating opportunities for steady employment at reasonable wages is the best way to tackle people out of poverty." (p. 1). "I argue that microcredit does yield some non-economic benefits (increased self-esteem and social cohesion, empowerment, less vulnerability), but that it does not significantly alleviate poverty." (p. 2). "While 'heart-warming case studies abound rigorous empirical analyses (on the impact of microcredit on poverty alleviation) are rare. A few studies have even found that microcredit has a negative impact on poverty; Poor households simply become poorer through the additional burden of debt. However, most studies suggest that microcredit is beneficial but only to a limited extent. The reality is less attractive than the promise." (p. 3-4).
Mosley and Hulme, 1998	"Microfinance has generated enormous enthusiasm among aid donors and NGOs as an instrument for reducing poverty in a manner that is financially self-sustaining." (p. 783). "Major comparative studies [...] on microfinance avoid calculations on poverty impact, often treating the fact that small loans are being made as in itself a proof that the poor are being reached and the fact that loans are being repaid as proof that incomes have increased. As a consequence we remain rather ignorant about the poverty impact of existing microfinance schemes." (p. 784).

Table 5-10: Citations on Microfinance and Poverty Alleviation

Reference	Citations on Bottom of the Pyramid
Citations on its Importance for Poverty Alleviations	
Olsen and Boxenbaum, 2009	"'The BOP concept proposes that there is a strong business case associated with the pursuit of the largely untapped purchasing power at the bottom of the world's economic pyramid. By viewing consumers in the developing world as resourceful entrepreneurs and value-conscious consumers rather than as victims, large MNCs can make significant profits and simultaneously help alleviate poverty by selling products to the poor." (p. 100-101).
Pitta, Guesalaga and Marshall, 2008	"The BOP approach to earning corporate profits has gained considerable attention in the marketing literature. It has awakened managers to the potential of serving an unserved market and alleviating the level of global poverty while still earning a profit." (p. 393).
Raufflet, Berranger and Aguilar-Platas, 2008	"In sum, this marketing-driven approach [=BOP] proposed two interrelated models of social change: 1) Business will alleviate poverty by creating products that the poor will be able to afford; doing so, the poor will become consumers of products previously financially unaffordable and physically non-accessible to them;

Reference	Citations on Bottom of the Pyramid
	2) Business will alleviate poverty through enterprise channel, the employment of poor for distribution of products to other poor" (p. 37).
Prahalad and Hart, 2002	"Treating the BOP as a market can lead to poverty reduction, particularly if NGOs and community groups can join with MNCs and local companies as business partners. The development of markets and effective business models at the BOP can transform the poverty alleviation task from one of constant struggle with subsidies and aid to entrepreneurship and the generation of wealth. When the poor at the BOP are treated as consumers, they can reap the benefits of respect, choice and self-esteem and have an opportunity to climb out of poverty." (p. 99). "I have no doubt that the elimination of poverty is possible by 2020" (p. 112).
Prahalad and Hammond, 2002	"By stimulating commerce and development at the BOP, MNCs could radically improve the lives of billions of people and help bring into being a more stable, less dangerous world. Achieving this goal does not require MNCs to spearhead global social development initiatives for charitable purposes. They need only act in their own self-interest for there are enormous business benefits to be gained by entering developing markets." (p. 4). "Big companies are not going to solve the economic ills of developing countries by themselves, of course. It will also take targeted financial aid from the developed world and improvements in the governance of the developing nations themselves. But it is clear that prosperity can come to the poorest regions only through the direct and sustained involvement of MNCs." (p. 5).
Citations on its Potential Contribution to Poverty Alleviations	
Walsh, 2005	"The BOP initiatives clearly improve the lives of the poor. The poor may not have more money to spend but they can better spend the money they do have. In so doing these initiatives help relieve the scourge of poverty. Of course, once MNCs target the poor and invited them to co-create these goods and services, the poor may be able to create incomes that were once inaccessible to them.." (p. 478). "In the end, we need to know much more about when and how BOP investments work to eliminate poverty. We also need to know how they may complement foreign aid, philanthropy and other kinds of anti-poverty initiatives.." (p. 480).
Bais, 2008	"There is growing evidence that a market-based approach on the BOP might work.." (p. 1).
Citations on its Potential Limitations for Poverty Alleviations	
Karnani, 2007b	"The popular BOP proposition argues that large companies can make a fortune by selling to poor people and simultaneously help eradicate poverty. This is, at best, a harmless illusion and potentially a dangerous delusion [...] We need to see the poor as producers, and emphasize buying from them, rather than selling to them. The only way to alleviate poverty is to raise the real income of the poor.." (p. 99). "'The BOP proposition is characterized by much hyperbole and very weak research methodology. The fortune and glory at the BOP is a mirage [...] The private sector can help alleviate poverty by focusing on the poor as producer. Certainly, the best way for private firms to help eradicate poverty is to invest in upgrading the skills and productivity of the poor and to help 109 create more employment opportunities for them. This is the win-win solution; this is the real fortune at the BOP." (p. 109-110).
Bais, 2008	"So far the results [of BOP] are mixed. Some attempts completely failed, others are still struggling, and some are likely to be successful.." (p. 3). "The most promising examples of BOP-products are those developed by

Reference	Citations on Bottom of the Pyramid
	local companies and local NGOs." (p. 15). "Little is known about the effective business strategies or the impact of existing projects [and] there is still much discussion about which criteria should be met to deserve the title 'BOP' project" (p. 16).
Karnani, 2007b	"The popular BOP proposition argues that large companies can make a fortune by selling to poor people and simultaneously help eradicate poverty. This is, at best, a harmless illusion and potentially a dangerous delusion [...] We need to see the poor as producers, and emphasize buying from them, rather than selling to them. The only way to alleviate poverty is to raise the real income of the poor.." (p. 99). "'The BOP proposition is characterized by much hyperbole and very weak research methodology. The fortune and glory at the BOP is a mirage [...] The private sector can help alleviate poverty by focusing on the poor as producer. Certainly, the best way for private firms to help eradicate poverty is to invest in upgrading the skills and productivity of the poor and to help 109 create more employment opportunities for them. This is the win-win solution; this is the real fortune at the BOP." (p. 109-110).
Bais, 2008	"So far the results [of BOP] are mixed. Some attempts completely failed, others are still struggling, and some are likely to be successful.." (p. 3). "The most promising examples of BOP-products are those developed by local companies and local NGOs." (p. 15). "Little is known about the effective business strategies or the impact of existing projects [and] there is still much discussion about which criteria should be met to deserve the title 'BOP' project" (p. 16).
Seelos and Mair, 2006	"The debate around the idea of viewing the poor as customers and not merely as the recipients of donations has been intriguing but we still lack (novel business) models for companies to realize this vision." (p. 2).
London, 2007	"While interest and debate about the BOP as a poverty alleviation perspective is growing, most of the current research has focused on business strategies for organizations interested in exploring these markets. Indeed, a deep exploration of the unique poverty alleviation implications of a BOP perspective has lagged" (p. 2). "BOP as producers and BOP as consumers are likely to have different poverty alleviation outcomes." (p. 12). The BOP perspective is not presented as a panacea that will replace or supersede other proven poverty alleviation approaches. Rather, it is better viewed as a perspective that can complement other efforts." (p. 28). "The role BOP can play in alleviating poverty will also vary depending on the context (external environment) and segments addressed (part of pyramid)" (p. 29).

Table 5-11: Citations on Bottom of the Pyramid and Poverty Alleviation

Reference	Citations on Social Entrepreneurship
Citations on its Importance for Poverty Alleviations	
Yunus, 2006	"It is time to move away from the narrow interpretation of capitalism and broaden the concept of market by giving full recognition to social business enterprises. Once this is done social business entrepreneurs can make the market work for social goals as efficiently as it does for private goals." (p. 5).
Citations on its Potential Contribution to Poverty Alleviations	
Smith and Feldman Barr, 2007	"SE involves the creation of innovative, sustainable solutions to immediate social problems with an emphasis on those who are marginalized or poor" (p. 27). "Consistent with research that suggests that the most effective

Reference	Citations on Social Entrepreneurship
	approach to reducing poverty is raising the income of the poor [the entrepreneurs in the sample] utilized SE as an innovative vehicle for long-term change to address poverty in the developing world […] These initiatives demonstrate the potential effectiveness of SE as a means to reducing poverty." (p. 28)
Yunus, 2010	"Social business has the potential to reverse this disparity [between the rich and the poor] because it addresses the poor directly and deliberately." (p. 203).
Easterly and Miesing, 2007	"Social ventures are innovative businesses that operate with social purposes to provide services to disadvantaged individuals or the community that the market does not […] They have been used to reduce poverty through job creation for the chronically unemployed, help impoverished communities produce their own products rather than importing them, create markets for products produced by impoverished communities and provide job training to help the chronically unemployed acquire employable skills." (p. 3). "Through enterprise development that creates local value and builds self-reliance among chronically unemployed, social ventures often have succeeded in reducing poverty in areas where charities have failed." (p. 4).
Zahra et al., 2009	"Social entrepreneurs make significant and diverse contributions to their communities and their societies, adopting business models to offer creative solutions to complex and persistent social problems" (p. 519). "Throughout the world, socially conscious individuals have introduced and applied innovative business models to address social problems previously overlooked by business, government and NGOs. These entrepreneurs have played a vital role in ameliorating adverse social conditions and have become visible agents of change in developing economies, where they have applied innovative and cost-effective methods to address nagging social problems (i. e. poverty) that have defied traditional solutions." (p. 520).
Seelos and Mair, 2004	"The efficiencies of markets combined with the resources and managerial expertise of large multinationals are considered crucial to tackling many of the world's development problems" (p. 1-2). "[Social entrepreneurs] come up with new solutions that are designed starting from local needs rather than the centralized assumptions of large institutions about what needs to be done." (p. 3) "Social entrepreneurs find new and efficient ways to create products and services that directly cater to social needs that remain unsatisfied by current economic and social institutions" (p. 4).
Citations on its Potential Limitations for Poverty Alleviations	
Mair and Schoen, 2005	"Although SE organizations have begun to receive more scholarly attention, we still know relatively little about *how* they are able to create both social and economic value." (p. 1).
Eder and Öz, 2008	"The number of entrepreneurs that eventually become successful in their business ventures is in fact not that high in any given society." (p. 124).
Raufflet, Berranger and Aguilar-Platas, 2008	"These approaches [of SE] are still in their infancy: they lack solid conceptual definitions and rigorous models of social change; these approaches would probably gain from systematic research that would evaluate their effects on poverty alleviation." (p. 54).
Austin, Stevenson and Wei-Skillern, 2006	"Mobilizing human and financial resources for SE is an extremely onerous task. […] Virtually all social issues require far more resources than any single organization is capable of mobilizing independently to solve. Networking across organizational boundaries to create social value is a power-

Reference	Citations on Social Entrepreneurship
	ful strategy for social entrepreneurs.“ (p. 17-18).
Seeloas and Mair, 2005	“To make a significant contribution to sustainable development social entrepreneurship must reach a critical mass of initiatives around the globe.“ (p. 244).

Table 5-12: Citations on Social Entrepreneurship and Poverty Alleviation

Appendix D: Theoretical Part – Literature Review Citations Section 2.4

Reference	Citations on Cross-Sector Partnerships
Citations on its Importance for Poverty Alleviations	
Singer, 2006	"Currently, entrepreneurs and corporations overwhelmingly do not view alleviation of global poverty as a strategic priority. Yet, business activity can have a negative as well as a positive effect on each distinctive form of poverty. In order to reduce poverty, entrepreneurs have to find ways of limiting the negative aspects. This might be achieved by deliberately augmenting strategies so that they can achieve a synthesis in partnership with governments and NGOs." (p. 225)
Raufflet, Berranger and Aguilar-Platas, 2008	"The starting point is the existence of a gap between different sections of society: While businesses – mainly MNCs – tend to ignore the poor, poverty alleviation organizations such as NGOs and governments, while they do not have the sufficient capacity to alleviate poverty without businesses, often ignore and mistrust businesses. The common challenge shared across society regarding poverty alleviation thus deals with the creation of a sense of interdependency between the sectors of society." (p. 36-37).
Selsky and Parker, 2005	"One type of collaborative engagement is partnership among business, government and civil society – the three main societal 849 sectors – that address social issues and causes. In cross-sector social-oriented partnerships organizations jointly address challenges such as [...] poverty alleviation." (p. 850).
Mert, 2009	"Recalling the objectives founded in the UN Millennium Declaration, particularly in regard to developing partnerships through the provision of greater opportunities to the private sector, NGOs and civil society in general so as to enable them to contribute to the realization of the goals and programs of the organization, in particular in the pursuit of development and the eradication of poverty. [...] Stressing that efforts to meet the challenges of globalization could benefit from enhanced cooperation between the UN and all relevant partners, in particular the private sector" (UN, 2001, p. 329).
Singer, 2006	"Poverty is revealed as a multi-dimensional construct, with poverty alleviation, in turn, requiring a multi-faceted approach. [...] Pro-poor economic growth involves a constellation of partnerships amongst entrepreneurs, governments and other institutional agencies, but with each partner intentionally pursuing multiple objectives that include poverty reduction." (p. 226).
Sharma, 2007	"Innovative financing mechanisms engaging diverse stakeholders for developing poverty reduction strategies have been engineered as possible neo-institutional structures to counter the world's most daunting problems" (p. 625). "Multi-stakeholder partnerships involving donors, governments, private sector and civil society have been presented as an all-inclusive institutional mechanism to affect change" (p. 626).
Item, 2007	"To succeed in sustainable development requires collaborative effort" (p. 13).
Ginsberg, 2002	"New partnerships could be a doorway through which we – government, NGOs and citizens – collectively move through the next decade pursuing the notion of sustainable development, creating improved quality of life and diminishing injustice, inequality and ecosystem degradation." (p. 471).

Reference	Citations on Cross-Sector Partnerships
Warhurst, 2005	"This paper suggests that the sustainable development challenge might best be tackled through a future characterized by a broad spectrum of partnerships between governments, development agencies, businesses, civil society and local communities." (p. 154). The MDGs, as compelling as they may be, however, involve an enormous task that cannot be achieved by any one single actor and that cannot be achieved without business, given the ten-fold increase in FDI over the past 10 years." (p. 155). "Organizations are increasingly working in partnership to address major societal issues that cannot be mitigated by any single actor or institution." (p. 163).
Curtis, 2006	"A common understanding [of poverty reduction strategies] at the present time is that governments lack the ability to go it alone [...]; hence the call for private sector involvement, for partnership and a role for civil society and a state that enables rather than provides." (p. 150).
Campbell, 2005	"Economic issues associated with poverty are complex and require holistic responses in order to realize the goals of SD. While business alone may have significant economic impacts, the link between business-level behavior and macro-level development aspirations is unclear. By developing a sound grasp of how companies understand and manage these impacts, we are better placed to understand how corporate responsibility clusters or partnerships with other companies, CSO and governments can harness corporate impacts to bring about more sustainable development at the macro level." (p. 413).
Wadham, 2009	"Partnership between business and NGOs provides one way of tackling global challenges like poverty." (p. 2).
Douglas, 2008	"'If we are to tackle global poverty, [...] the solution lies not with the government alone, but in partnership of voluntary, state, faith and private institutions and citizens" (p. 48). "Aid is a necessary part of the solution to global poverty, it will not be sufficient to end poverty." (p. 49).
Stefanovic, 2007	"A number of examples show that creative partnerships and approaches to reducing poverty through trade may be even more important than resources" (p. 6).
Gold, 2004	"Partnerships (between the business community, existing NGOs and global community networks) present innovative strategies for [...] effectively eradicating poverty." (p. 636).
Thavisin, 2000	"To advance social development as an integral part of national and international policy and strategies, all governments should bear in mind the importance of partnerships in addressing the problems of poverty alleviation." (p. 98).
Oketch, 2004	"Social partnerships between business, government and NGOs are increasingly being developed throughout the world as potential answers to the general problem of reconciling the economic and social dimension of human activity." (p. 11).
Citations on its Potential Contribution to Poverty Alleviations	
Humphrey, 2006	"The analysis has highlighted areas where development partnerships can contribute to meeting the challenges to sustaining growth and poverty reduction. [...] The goal must be to work towards identifying how new partnerships for development between the public and private sectors and between Asian countries and the broader development community can contribute to improving the welfare of the poor in Asia by making sustained growth work more effectively for the poor." (p. 47-48).

Reference	Citations on Cross-Sector Partnerships
Karher, Iyani and Shannon, 2007	"Effective action (with regard to poverty alleviation) must involve local and global stakeholders in collaboration." (p. 69). "... a partnership of MNCs, international development agencies and NGOs." (p. 75). "What is necessary (for poverty alleviation) are multiple strategies that involve a high level of collaboration and coordination with all organizations involved in the development process. Grounded in the active participation of those in poverty, this kind of coordinated approach will maximize resources, end duplications of services, and carefully target areas of need." (p. 83).
Spielman, Hartwich and von Grebmer, 2007	"Public-Private-Partnerships (PPPs) [...] in development are increasingly viewed as effective means of conducting advanced research, developing new technologies and developing new products for the benefit of small-scale, resource-poor farmers and other marginalized groups in developing countries" (p. 1). "We ask whether PPPs are effectively targeting the poor in developing countries. This study suggests that PPPs are serving a wide variety of research objectives. [...] Partnership with the private sector is still at a very nascent stage. [...] Few provide effective analysis of their poverty-targeting strategies. [...] However, it is important to note that a one-size-fits-all-approach to PPPs is counterproductive." (p. 5-6).
Kanji, 2004	"[This paper] demonstrates that collaboration between government, companies and CSO at the national level can contribute to [...] sustainable development." (p. 82). "This case illustrates the potential for collaboration between government, NGOs and employees for improvements in wages and working conditions in one location." (p. 86).
Dolan and Scott, 2009	"Increasing numbers of corporations are trying to capture one of the largest untapped consumer markets – the world's poor – in ways that are not only economically profitable but socially responsible. One type of initiative that has gained increased traction is trading partnerships between MNCs and women's informal exchange networks, creating micro-enterprise opportunities that not only deliver soap and mobile phones, but financial empowerment for women." (p. 203).
Rein and Stott, 2009	"Spreading rapidly across contexts and fields of intervention, cross-sector social partnerships (CSSPs) are now at the forefront of creative organizational models to offer innovative solutions to complex and persistent social problems, balancing nonprofit attitude to social service with business entrepreneurial orientation. To this end they provide unusual combinations of complementary core competencies to create more inclusive-participatory models aimed at strategically addressing social burdens in sectors, such as [...] poverty alleviation. [...] CSSPs have started to be perceived as strategically better responses to a changed and challenging macro-situation made of raising costs and decreased public and private grants and donations. [...] CSSPs are interpreted as new strategies of engagement with communities that are both more effective in their ability to pursue social impact and more relevant to the company core value proposition. As a result CSSPs have become the new organizational zeitgeist in dealing with societal issues. [...] Literature has shared a tendency to portray these forms of collaboration as a 'kind of magic bullet capable of providing solutions to diverse development problems across a variety of settings through win-win situations where all stakeholders benefit' (p. 80)".

Reference	Citations on Cross-Sector Partnerships
Coxon and Munce, 2008	"In recent years the international development community's concern to redress the apparent failure of half a century of development efforts has given rise to a number of initiatives aimed at […] reducing poverty worldwide. One outcome of these initiatives has been a refocusing […] on enhanced partnership." (p. 147). "We would suggest that the ingredients in this particular partnership […] may provide a promising, if modest, model for gender equitable poverty reduction in developing countries (p. 215).
Bishop, 2007	"The most likely (future scenario on poverty alleviation approaches) is a continuum of the trend for flexible, ad hoc partnerships, issue by issue, between different players. That may well be the most effective way forward, as well as the most feasible – given the resistance of multilateral institutions to serious reform." (p. 1).
Lapreye, 2011	"[CSPs] could efficiently help to tackle multi-dimensional aspects of poverty and achieve MDG 1 by 2015 (through their benefits of job creation, local empowerment and community income creation)" (p. 233).
Cycyota and Volkland, 2007	"The enormous scope of poverty makes reliable solutions difficult to imagine and more difficult to implement. […] We suggest that the issue may be effectively, if slowly, addressed by an approach that begins with individual businesses and builds up to a level that involves governments NGOs and MNCs. The end result is an innovative and sustainable manner that harnesses the resources and abilities of multiple levels of organizations." (p. 44).
Vermeulen, Navir and Mayers, 2008	"At the Earth Summit in Johannesburg in 2002, partnerships were touted as one of the key routes to sustainable development. [...] Positive local impacts of company-community deals include sharing of risks, better returns to land than otherwise possible, opportunities for income diversification, access to paid employment, development of new skills, upgrading of local infrastructure and environmental improvement. However, company-community deals have not yet proved sufficient to lift people out of poverty. They remain supplementary rather than central to income generation." (p. 1).
Goldsmith, 2011	"A growing movement commends a targeted business-oriented approach as the best chance to reduce global poverty. […] This appealing vision may underplay the importance of multi-stakeholder collaboration to make it work. […] A review of previously documented cases shows the market-oriented approach can reach low-income and socially excluded house-holds, but the payoff usually depends on cross-sector subsidies from non-profit or government organizations. […] My literature review suggests that poverty-fighting commercial enterprises are usually helped by charitable or public organizations. That unremunerated help, whether monetary or in kind, appears to be critical to success on the double bottom line." (p. 15-16).
Gardetti, 2005	"People are slowly beginning to understand that poverty alleviation re-quired complex collaborative alliances. Through their knowledge about the needs of people under conditions of poverty and the culture of the poor, these organizations can promote and provide for the business-poverty relationship, thus resulting in social inclusion, development and increased opportunities." (p. 70).

Citations on its Potential Limitations for Poverty Alleviations	
Andrews and Entwistle, 2010	"Cross-sectoral partnerships are increasingly seen as a solution to the most pressing social problems facing contemporary societies. Sectoral rationales for partnership suggest that public, private and nonprofit organizations each possess distinct advantages that can enhance the effectiveness, efficiency and equity of public agencies' efforts to address social issues. [...] The results indicate that public-public partnership is positively associated with effectiveness, efficiency and equity but that public-private partnerships are negatively associated with effectiveness and equity. Public-nonprofit partnership is unrelated to performance. Our study therefore suggests that cross-sectoral partnership does deliver but that the prospects of public service improvement may depend on the sectoral choice that organizations make." (p. 679).
Franceys and Weitz, 2003	"A very limited number of public private community partnerships with international operations were found to be dramatically improving services to some of the poor." (p. 1083). "PPPs are usually by default rather than design delivering some limited benefits to the poor." (p. 1092).
Brinkerhoff and Brinkerhoff, 2011	"PPPs have long been advocated and analyzed as organizational solutions to pressing social problems that call for the comparative advantages of government, business and civil society. However, ongoing questions remain about how to design, manage and assess PPPs. The large literature on PPPs suffers from conceptual imprecision and is weakly integrated." (p. 2). "Despite their originating rationale, many PPPs may not achieve their intended public benefits, due to poor implementation or skewed incentives." (p. 9).
Mahantry et al., 2009	"The term partnership has become a watch-word for development organizations that aim to mobilize the resources and collaboration needed to achieve long-term goals such as poverty reduction. Achieving effective collaboration in practice, however, can be challenging. Key findings include the observation that the role of individuals in maintaining partnerships often goes unrecognized and needs to be supported in appropriate ways; clearly defined and focused areas of collaboration are essential; a formal basis for the partnership needs to be backed with strong informal communication and collaboration processes; and while partners bring distinctive knowledge and networks to a partnership, some degree of evenness in the scale and type of resources committed to the partnership is important." (p. 859).
Gold, 2004	"Preliminary research, however, has been inconclusive as to the positive impact that such joint ventures and partnerships (between state and non-state actors) can have on poverty alleviation in developing countries." (p. 635).
Rein and Stott, 2009	"Partnership has been promoted by large numbers of corporations, governments, international agencies and NGOs as the most effective way of working towards the achievement of sustainable development [...] Although some work has been undertaken in the study of partnerships, there is not yet a significant body of critical analysis on their impact." (p. 79).

Table 5-13: Citations on Cross-Sector Partnerships and Poverty Alleviation

Reference	Citations on the Combinations of Cross-Sector Partnerships with the Business Strategies on Poverty Alleviation
Citations on the Combination of CSPs and Business-as-Usual	
Gifford, Kestler and Anand, 2010	"When entering developing nation markets, MNCs may need to expand dramatically the potential field of alliance partners to meet the challenges of gaining needed resources and expertise. These additional partners are often non-traditional and can include NGOs, community groups and even village and local tribal governments. […] These partners provide important information and contacts in the local community and business environment that are not generally available in the corporate sector." (p. 305-306).
Ayanwu, 2006	"New and effective investment opportunities […] now encourage PPP which declined investment risks for FDI in Africa" (p. 42).
Citations on the Combination of CSPs and Corporate Philanthropy	
Tang and Li, 2009	"Companies engage in strategic philanthropy through long-term practices such as establishing partnership with governments, NGOs and universities, or establishing awards or grants on an on-going basis to encourage societal participation in causes important to them (i. e. poverty)." (p. 208).
Moran, 2008	"Perhaps the most noteworthy aspect of the emerging aid consensus has been increased collaboration between the sectors through development-driven PPPs. The catalyst for many PPPs has been private philanthropic foundations." (p. 26).
Citations on the Combination of CSPs and Microfinance	
UNDP, 2004	"Facilitating cooperation and partnership between public and private players to enhance access to such key factors as financing, skills and basic education" (p. 2).
Pitta, Guesalaga and Marshall, 2008	"To reduce the cost of capital, (for microloan provisions) perhaps collaboration with funding sources like the World Bank or other NGOs will be necessary." (p. 398).
Rowe, Gavrilova, Velev and Shaw, 2009	"Microfinance is a tool that can be used to empower marginalized households […] Long-term success, however, depends on intentional partnerships with other CS and government actors who can influence social change" (p. 235).
Citations on the Combination of CSPs and Bottom of the Pyramid	
Prahalad and Hammond, 2002	"Succeeding in BOP markets requires companies to think creatively. […] In addition to limiting risks for each player, partnerships maximize the existing infrastructure […] MNCs seeking partners should look beyond businesses to NGOs and community groups. They are key sources of knowledge about customers' behavior, and they often experiment most with new services and delivery models." (p. 10).
London, 2007	"The BOP perspective is premised on the BOP venture establishing a set of mutually beneficial partnerships with local organizations, community institutions and entrepreneurs currently operating at the BOP" (p. 19).
Bais, 2008	A shift in BOP initiatives occurred from selling to co-creating and involvement of the poor: "This strategy of co-creation requires unconventional partnerships between companies, NGOs and government bodies" (p. 3). "Commercial market development efforts thus are combined with traditional development tools like community organization, consumer education, stimulating local entrepreneurship, job creation, micro leans and cultural sensitivity. It also asks for government participation, not just to enhance the economic activity by providing legislation, land rights, transportation, airfields or landing strips." (p. 10).

Reference	Citations on the Combinations of Cross-Sector Partnerships with the Business Strategies on Poverty Alleviation
Olsen and Boxenbaum, 2009	"MNC in collaboration with other global actors are considered instrumental in achieving these goals (of addressing the BOP)" (p. 103).
Walsh, 2005	"Prahalad wants to celebrate public-private collaboration and nonprofit initiatives and claim microfinance and social entrepreneurship activities as part of a business momentum for the BOP phenomenon." (p. 477).
Kelley, Werhane and Hartman, 2008	"Various sectors can work together (in engaging at the BOP) in ways that are beneficial to all, including global companies. Whatever the nature of the role, it is clear that collaboration is essential" (p. 26).
Gardetti, 2005	"The Base of the Pyramid offers opportunities to create value for all parties, as well as opportunities for both sustainable and human development. [...] This requires the use of the imagination to forge alliances and make cooperative approaches that blend the capabilities of business with those of both CS and the public sector in order to speed up development, even in the poorer regions" (p. 75).
Pitta, Guesalaga and Marshall, 2008	"There is recognition that serving the BOP requires the involvement of multiple players, including private companies, governments, NGOs, financial institutions and other organizations e.g. communities." (p. 398).
Newell and Frynas, 2007	"Partnerships with NGOs, development agencies and local communities are said to be able to help private firms to develop new markets, while providing the poor with access to markets and services." (p. 670).
Citations on the Combination of CSPs and Social Entrepreneurship	
Easterly and Miesing, 2007	"Social ventures can take numerous forms and have been established under varying legal and operating structures, [including] partnerships between public, private or not-for-profit sector organizations." (p. 6).
Van Slyke and Newman, 2006	"A social entrepreneur is one who 'develops a strategic service vision, a competitive strategy, a strategy for building networks and partnerships, leads, retains and rewards people, manages entrepreneurially, treats donors as investors, works with different communities, develops viable earned income strategies, considers the scale of the project and strategies for success, and is able to manage organizational change" (Dees, 2001, p. 348).
Yunus, Moingeon and Lehmann-Ortega, 2010	"The second step in building social business models is to leverage expertise and resources by setting up partnerships. [....] The main advantage of collaborative agreements lies in the pooling of resources and knowledge leveraged by the partners, which may in turn lead to the development of a broader portfolio of resources for firms in the network. Cooperation is considered as a major factor of success for pro-active CSR strategies and research stresses the importance of long-term relationships among such actors" (p. 314). "Such partnerships between business and non-for-profit organizations can be highly productive and low in risk, as they take place between actors who are not in direct competition with each other." (p. 315).

Table 5-14: Citations on Cross-Sector Partnerships in Combination with Business Strategies on Poverty Alleviation

Reference	Citations on the Combinations of the Business Strategies on Poverty Alleviation
Citations on the Combination of Corporate Philanthropy and Bottom of the Pyramid	
Bishop, 2007	"The final group of new players [in fighting global poverty] is multinational companies, which increasingly are getting involved in development and poverty issues. They are doing so in several guises, including through corporate philanthropy, CSR, global supply chain management and BOP strategies to sell to poorer consumers. Of these, the latter two are most likely to deliver sustained impact, as they are core to the firm's profit-generation strategy.." (p. 8).
Citations on the Combination of Corporate Philanthropy and Social Entrepreneurship	
Goldsmith, 2011	"Social venture capital: Typical sources are donor agencies, foundations and increasingly a new breed of private equity devoted to the public good – known as social venture capital or impact investing.." (p. 20).
Citations on the Combination of Microfinance and Bottom of the Pyramid	
Rowe, Gavrilova, Velev and Shaw, 2009	"Microfinance practioneers need to be willing to focus on the BOP and need on-going measurement systems of poverty outreach [...]. [Microfinance] must engage other key actors through strategic and collaborative alliances since microfinance alone cannot address social behavioral changes.." (p. 245)
Walsh, Kress and Beyerchen, 2005	"Prahalad wants to celebrate public-private collaboration and nonprofit initiatives and claim microfinance and social entrepreneurship activities as part of a business momentum for the BOP phenomenon.." (p. 477).
Citations on the Combination of Microfinance and Social Entrepreneurship	
Goldsmith, 2011	"'The first variety of social enterprise makes credit, savings, and insurance available to low-income clients (often female) in the informal economy. These are people who are excluded from mainstream trade and industry and who must rely largely on themselves to survive. [...] Traditional savings and credit groups fail to mobilize sufficient capital for these small-scale business people to break out of the poverty trap. Microfinance intends to fill that gap." (p. 17-18).
Citations on the Combination of Bottom of the Pyramid and Social Entrepreneurship	
Goldsmith, 2011	"'The second sub-type of social enterprise sells useful products other than financial services to 'the Base of the Pyramid' . [...] Often such pro-poor marketing schemes are carried out by a special social venture unit of a MNC – with deep corporate pockets to back it up." (p. 18).
Seelos and Mair, 2006	Recommendations for companies wishing to enter BOP markets: "Study the initiative of *social entrepreneurs* and challenge paradigms about poverty-related problems and solutions.." (p. 12).

Table 5-15: Citations on the Combination of Business Strategies on Poverty Alleviation

Appendix E: Theoretical Part – Overview Tables per Business Approach

Business strategy	Business-as-Usual
Organizational characteristics and activities	
Possible organizational type	Business
Possible company size	MNCs or SMEs (focus on MNCs operating in developing countries)
Possible sectors	All
Part of core activities of firm	FDI yes but poverty as an issue not
Activities undertaken	None with regard to poverty; solely regular business activities
Underlying business notion	
Potential motivation	Profit maximizations; shareholder view
Business benefit	Financial (business growth and revenue generation via FDI)
Business self-sufficiency	Yes
Scalability	High if profitable FDI opportunities can be obtained
Compatibility	
Possibility for partnership	Yes, but more likely with companies and not related to poverty
Possibility for combination with other strategies	Low due to lack of focus on CSR and poverty alleviation
Impact on poverty alleviation	
Poverty alleviation approach	Mainly indirect (via economic growth); top-down
Poverty dimension addressed	Economic but no active focus on tackling poverty
View of poor as	Passive members of communities
Involvement of poor	Low; poor are not directly addressed or involved
Possible benefits for the poor and their communities	Financial (potential employment and income; economic growth; tax payments etc.) and non-financial (skill gains; enhanced quality of life etc.)
Evaluation	
Main criticism	Insufficient as it does not directly address poverty → Need for companies to go beyond to proactively combat poverty; Crowding out of local firms, race to the bottom, worse working conditions, dependency
Challenge and limitations	Economic growth is not enough to permanently life people out of poverty as other poverty dimensions remain neglected in practice; benefits on poor are marginal due to inequality and social exclusion
Sustainability	Low due to dependence and lack of inclusion of the poor
Effectiveness	Limited due to prime focus on effects of economic growth
Classification	Inactive

Table 5-16: Overview Table for Business-as-Usual

Business strategy	Classical Corporate Philanthropy	Strategic Corporate Philanthropy
Organizational characteristics and activities		
Possible organizational type	Business	Business
Possible company size	Mostly MNCs	Mostly MNCs
Possible sectors	All	All
Part of core activities of firm	No	Partially
Activities undertaken	Solely financial donations to external NGOs or governmental organizations	Financial and non-financial assistance via own foundation, training, employee giving and volunteering
Underlying business notion		
Potential motivation	External pressure	Strategic expertise
Business benefit	Non-financial (improved stakeholder relations)	Financial and non-financial (strategic business benefits; improved stakeholder relations)
Business self-sufficiency	No	Partially
Scalability	Low; restricted on funds	Rather low; restricted on funds and strategic opportunities
Compatibility		
Possibility for partnership	Moderate with NGOs or governments; low with the poor; frequently limited in extensiveness	High with NGOs or governments; low with the poor; frequently more extensive
Possibility for combination with other strategies	Moderate with Microfinance, BOP and SE	High with Microfinance, BOP and SE
Impact on poverty alleviation		
Poverty alleviation approach	Direct and indirect; top-down	Direct and indirect; top-down
Poverty dimension addressed	All (depends on project)	All (depends on project)
View of poor as	Passive recipients of aid	Passive recipients of aid
Involvement of poor	Low	Low
Possible benefits for the poor and their communities	Improvements in various financial and non-financial poverty dimensions (differs)	Improvements in various financial and non-financial poverty dimensions (differs)
Evaluation		
Main criticism	Short-term, instable aid/ quick fix; low involvement of poor in process and creation of aid dependence, lack of new skills generation unrelated to business expertise and frequent outsourcing of projects to NGOs	Short-term, instable aid/ quick fix; low involvement of poor in process and creation of aid dependence, lack of new skills generation

Business strategy	Classical Corporate Philanthropy	Strategic Corporate Philanthropy
Challenge and limitations	Scalability restricted by funding; low self-sufficiency	Scalability restricted by funding; moderate self-sufficiency
Sustainability	Low	Low
Effectiveness	Low	Moderate
Classification	Reactive	Active

Table 5-17: Overview Table for Corporate Philanthropy

Business strategy	Narrow Microfinance	Broad Microfinance
Organizational characteristics and activities		
Possible organizational type	Business or NGO	Business
Possible company size	MNCs or SMEs	MNCs or SMEs
Possible sectors	All sectors possible but foremost financial	All sectors possible but foremost financial
Part of core activities of firm	Generally no	Generally yes
Activities undertaken	Provision of microcredit services	Provision of microcredit and further microfinance services; training and support services
Underlying business notion		
Potential motivation	Improved stakeholder relations; societal benefit	Potential business case; societal benefit
Business benefit	Mainly non-financial benefits (reputation, stakeholder relations etc.)	Financial and non-financial (reputation, stakeholder relations etc.)
Business self-sufficiency	Generally no (operating more like a charity)	Yes (operating as business entity)
Scalability	Moderate due to lack of self-sufficiency	High due to self-sufficiency
Compatibility		
Possibility for cooperation	Moderate with NGOs, governments or the poor; frequently limited in extensiveness	High with NGOs, governments or the poor; frequently more extensive
Possibility for combination with other strategies	Moderate with CP, BOP or SE	High with CP, BOP or SE
Impact on poverty alleviation		
Poverty alleviation approach	Indirect; Bottom-up	Indirect; Bottom-up
Poverty dimension addressed	Mainly economic	Economic and non-economic
View of poor as	Capable entrepreneurs	Capable entrepreneurs
Involvement of poor	High	High

Business strategy	Narrow Microfinance	Broad Microfinance
Possible benefits for the poor and their communities	Financial (access to credit which can be used for generation of income through entrepreneurial activities); non-financial (reduced vulnerability)	Same as for narrow microfinance plus skill enhancement through training and capacity building (non-financial)
Evaluation		
Main criticism	Credit not only given for income-related purposes; lack of entrepreneurial skills; need for training; credit may increase poverty through debt	Access to credit and training may improve living situation but might be insufficient for overall poverty alleviation
Challenge and limitations	Restriction on successful possible entrepreneurial start-ups; need for more stable employment opportunities	Restriction on successful possible entrepreneurial start-ups; need for more stable employment opportunities
Sustainability	Solely if entrepreneurial business models are sustainable	Solely if entrepreneurial business models are sustainable; gained skills increase social capital over the long run
Effectiveness	Mixed results; overall moderate impact on poverty alleviation; improvement of economic situation but insufficient on its own for a permanent lift out of poverty	Similar as for narrow approach but results are slightly more positive due to higher levels of acquired skills of poor
Classification	Reactive	Active

Table 5-18: Overview Table for Microfinance

Business strategy	Narrow Bottom of the Pyramid	Inclusive Bottom of the Pyramid
Organizational characteristics and activities		
Possible organizational type	Business	Business
Possible company size	All but mainly MNCs	All but mainly MNCs
Possible sectors	All	All
Part of core activities of firm	Generally no	Generally yes
Activities undertaken	Sale of goods to poor	Inclusion of poor in value chain
Underlying business notion		
Potential motivation	Business case	Business case Potential for value chain reconfiguration and need for skilled, local employees
Business benefit	Financial (business growth)	Financial (growth; cost reduction; favorable value chain configurations) and non-financial (skilled human resources)
Business self-sufficiency	High	High

Business strategy	Narrow Bottom of the Pyramid	Inclusive Bottom of the Pyramid
Scalability	High as long as profitable business cases can be found	High as long as profitable business cases can be found
Compatibility		
Possibility for partnership	Moderate with NGOs, governments or the poor; frequently limited in extensiveness	High with NGOs, governments or the poor; frequently more extensive
Possibility for combination with other strategies	Moderate with CP, Microfinance and BOP	High with CP, Microfinance and BOP
Impact on poverty alleviation		
Poverty alleviation approach	Mainly indirect; bottom-up	Mainly indirect; bottom-up
Poverty dimension addressed	Economic and social	Economic and social
View of poor as	Neglected consumers	Consumers, skilled producers, entrepreneurs and employees
Involvement of poor	Moderate to high	Very high
Possible benefits for the poor and their communities	Non-financial (Addressing of neglected needs and improved standards of living)	Financial (employment) and non-financial (social inclusion, empowerment, capacity building; improved standards of living)
Evaluation		
Main criticism	Better product choices are insufficient to end poverty; crowding out of local firms; need to provide long-term employment to poor; Exclusion of the poorest of the poor	Need to ensure that poorest of the poor are addressed and that social inclusion is in a long-term manner
Challenge and limitations	Limits of profitable business cases at the BOP	Limits of profitable business cases at the BOP and lack of trained employees/ entrepreneurs
Sustainability	High	High
Effectiveness	Rather high	High
Classification	Active	Proactive

Table 5-19: Overview Table for Bottom of the Pyramid

Business strategy	Narrow Social Entrepreneurship	Broad Social Entrepreneurship
Organizational characteristics and activities		
Possible organizational type	Business or NGO	Business
Possible company size	MNCs and SMEs	MNCs and SMEs
Possible sectors	All	All
Part of core activities of firm	Generally no	Generally yes
Activities undertaken	Sole support of external SE	Own engagement into SE activi-

Business strategy	Narrow Social Entrepreneurship	Broad Social Entrepreneurship
	organizations	ties related to poverty
Underlying business notion		
Potential motivation	Social and financial goals	Social goals
Business benefit	Creation of stakeholder value and addressing of societal needs	Creation of stakeholder value and addressing of societal needs; potential for financial value creation as well
Business self-sufficiency	Moderate (support via external funds possible)	High
Scalability	Moderate	High as long as business cases can be discovered
Compatibility		
Possibility for partnership	Moderate with NGOs, governments or the poor; frequently limited in extensiveness	High with NGOs, governments or the poor; frequently more extensive
Possibility for combination with other strategies	Moderate with CP, BOP and Microfinance	High with CP, BOP and Microfinance
Impact on poverty alleviation		
Poverty alleviation approach	Direct and indirect; bottom-up	Direct and indirect; bottom-up
Poverty dimension addressed	All (depends on initiative)	All (depends on initiative)
View of poor as	Skilled individuals which can be nurtured to become independent	Skilled individuals which can be nurtured to become independent
Involvement of poor	High	High
Possible benefits for the poor and their communities	Dependent on project but generally range from financial (employment and income) to non-financial (empowerment, social inclusion, capacity building, more favorable environmental conditions)	Dependent on project but generally range from financial (employment and income) to non-financial (empowerment, social inclusion, capacity building, more favorable environmental conditions)
Evaluation		
Main criticism	Uniqueness and diversity of approaches challenge scalability and comparison; need for willing social entrepreneurs; social goals solely as by-product	Uniqueness and diversity of approaches challenge scalability and comparison; need for willing social entrepreneurs; social exclusion via project selection
Challenge and limitations	Scalability limitations due to dependence on external funds	Limits of self-sustaining cases which involve the poor
Sustainability	Moderate	High
Effectiveness	Rather high	High
Classification	Active	Proactive

Table 5-20: Overview Table for Social Entrepreneurship

Business strategy	Cross-Sector Partnerships
Organizational characteristics and activities	
Possible company size	All
Possible sectors	All
Part of core activities of firm	Yes
Activities undertaken	Contingent on CSP approach in terms of number of initiatives, amount of partners, duration, scope and form of engagement
Underlying business notion	
Potential motivation	Increased efficiency; access to compatible resources; networking
Business benefit	Improved stakeholder relations; Access new resources
Business self-sufficiency	High
Scalability	High (if willingness of actors to cooperate)
Partnering approach	
Possible partners	NGOs, governments as well as the poor and their communities
Role of other societal actors	Sources of local knowledge; networker; providers of distinct expertise and resources
Compatibility	
Possibility for combination with business strategies	High with CP, Microfinance, BOP and SE; Low with Business-as-Usual
Impact on poverty alleviation	
Poverty alleviation approach	Direct and indirect; mostly bottom-up
Poverty dimension addressed	All (dependent on project)
View of poor as	Skilled community members
Involvement of poor	High
Possible benefits for the poor and their communities	Diversity of poverty dimensions addressed; high involvement and empowerment of poor; diversity of experts which increases sustainability of solutions
Evaluation	
Main criticism	Complexity of undertakings
Challenge and limitations	Success rate is contingent on the manner of implementation
Sustainability	High
Effectiveness	High
Classification	Contingent on partnership approach employed

Table 5-21: Overview Table for Cross-Sector Partnerships

Appendix F: Practical Part – Methodology

Sample Overview

Rank	Company	Country	Region	Sector	Classification
1	Wal-Mart Stores	United States	North and South America	Retail	Inactive
2	Royal Dutch Shell	The Netherlands	North and South America	Oil & Gas	Inactive
3	Exxon Mobil	United States	North and South America	Oil & Gas	Active
4	BP	United Kingdom	Europe	Oil & Gas	Reactive
5	Sinopec Group	China	North and South America	Oil & Gas	Inactive
6	China National Petroleum	China	Europe	Oil & Gas	Inactive
7	State Grid	China	North and South America	Utilities	Inactive
8	Toyota Motor	Japan	Europe	Automobiles & Parts	Inactive
9	Japan Post Holdings	Japan	Europe	Industrial Goods & Services (Other)	Inactive
10	Chevron	United States	Asia	Oil & Gas	Reactive
11	Total	France	Europe	Oil & Gas	Reactive
12	ConocoPhillips	United States	Europe	Oil & Gas	Inactive
13	Volkswagen	Germany	North and South America	Automobiles & Parts	Inactive
14	AXA	French	Asia	Financials	Reactive
15	Fannie Mae	United States	North and South America	Financials	Inactive
16	General Electric	United States	Europe	Technology	Active
17	ING Group	The Netherlands	North and South America	Financials	Reactive
18	Glencore International	Switzerland	North and South America	Basic Resources (Other)	Inactive
19	Berkshire Hathaway	United States	North and South America	Financials	Inactive
20	General Motors	United States	Asia	Automobiles & Parts	Inactive
21	Bank of America Corporation	United States	Asia	Financials	Reactive
22	Samsung Electronics	South Korea	Asia	Technology	Inactive
23	ENI	Italy	Asia	Oil & Gas	Reactive
24	Daimler	Germany	North and South America	Automobiles & Parts	Reactive
25	Ford Motor	United States	Europe	Automobiles & Parts	Inactive
26	BNP Paribas	France	Europe	Financials	Active
27	Allianz Group	Germany	Asia	Financials	Reactive

28	Hewlett-Packard	United States	Europe	Technology	Active
29	E.ON	Germany	Europe	Utilities	Inactive
30	AT&T	United States	Europe	Telecommunications	Inactive
31	Nippon Telegraph & Telephone	Japan	Asia	Telecommunications	Inactive
32	Carrefour	France	North and South America	Retail	Reactive
33	Assicurazioni Generali	Italy	Europe	Financials	Reactive
34	Petrobras	Brazil	Europe	Oil & Gas	Reactive
35	Gazprom	Russian Federation	Europe	Utilities	Inactive
36	J.P. Morgan Chase & Co.	United States	North and South America	Financials	Proactive
37	McKesson	United States	North and South America	Retail	Inactive
38	GDF Suez	France	North and South America	Oil & Gas	Active
39	Citigroup	United States	Asia	Financials	Reactive
40	Hitachi	Japan	North and South America	Technology	Inactive
41	Verizon Communications	United States	Europe	Telecommunications	Inactive
42	Nestlé	Switzerland	Europe	Food & Beverage (Other)	Reactive
43	Crédit Agricole	France	Asia	Financials	Active
44	American International Group	United States	Europe	Financials	Inactive
45	Honda Motor	Japan	North and South America	Automobiles & Parts	Inactive
46	HSBC Holdings	United Kingdom	North and South America	Financials	Reactive
47	Siemens	Germany	Europe	Technology	Reactive
48	Nissan Motor	Japan	Europe	Automobiles & Parts	Inactive
49	Pemex	Mexico	Europe	Oil & Gas	Inactive
50	Panasonic	Japan	Europe	Technology	Reactive
51	Banco Santander	Spain	Asia	Financials	Reactive
52	International Business Machines	United States	Asia	Technology	Inactive
53	Cardinal Health	United States	Europe	Retail	Inactive
54	Freddie Mac	United States	North and South America	Financials	Reactive
55	Hyundai Motor	South Korea	North and South America	Automobiles & Parts	Inactive

56	Enel	Italy	Asia	Utilities	Inactive
57	CVS Caremark	United States	Europe	Retail	Inactive
58	JX Holdings	Japan	Europe	Conglomerate (Other)	Inactive
59	Lloyds Banking Group	United Kingdom	Asia	Financials	Inactive
60	Hon Hai Precision Industry	Taiwan	Asia	Technology	Inactive
61	Tesco	United Kingdom	North and South America	Retail	Reactive
62	United Health Group	United States	Asia	Health Care (Other)	Inactive
63	Wells Fargo	United States	Europe	Financials	Reactive
64	Aviva	United Kingdom	Europe	Financials	Reactive
65	Metro Group	Germany	North and South America	Retail	Inactive
66	PDVSA	Venezuela	North and South America	Oil & Gas	Inactive
67	Statoil	Norway	Europe	Oil & Gas	Inactive
68	Électricité de France	France	Asia	Utilities	Inactive
69	Lukoil	Russian Federation	Asia	Oil & Gas	Reactive
70	Valero Energy	United States	Europe	Oil & Gas	Inactive
71	BASF	Germany	Europe	Chemicals (Other)	Reactive
72	Société Générale	France	Asia	Financials	Reactive
73	Sony	Japan	Asia	Technology	Reactive
74	ArcelorMittal	Luxembourg	Europe	Basic Resources (Other)	Inactive
75	Deutsche Telekom	Germany	North and South America	Telecommunications	Inactive
76	Kroger	United States	Asia	Retail	Inactive
77	Industrial and Commercial Bank of China	China	North and South America	Financials	Inactive
78	Telefónica	Spain	Europe	Telecommunications	Reactive
79	BMW	Germany	Asia	Automobiles & Parts	Reactive
80	Procter & Gamble	United States	Asia	Personal & Household Goods (Other)	Inactive
81	Nippon Life Insurance	Japan	Europe	Financials	Inactive
82	SK Holdings	South Korea	Europe	Oil & Gas	Inactive

83	EXOR Group	Italy	Europe	Financials	Inactive
84	AmerisourceBergen	United States	Europe	Retail	Inactive
85	Costco Wholesale	United States	Europe	Retail	Inactive
86	Petronas	Malaysia	Europe	Oil & Gas	Inactive
87	China Mobile Communications	China	North and South America	Telecommunications	Inactive
88	Munich Re Group	Germany	Europe	Financials	Reactive
89	Toshiba	Japan	Europe	Technology	Inactive
90	Peugeot	France	North and South America	Automobiles & Parts	Inactive
91	Prudential	United Kingdom	Europe	Financials	Inactive
92	Vodafone	United Kingdom	Europe	Telecommunications	Reactive
93	Deutsche Post	Germany	North and South America	Industrial Goods & Services (Other)	Active
94	Repsol YPF	Spain	Europe	Oil & Gas	Reactive
95	China Railway Group	China	North and South America	Industrial Goods & Services (Other)	Inactive
96	Dexia Group	Belgium	Asia	Financials	Reactive
97	Groupe BPCE	France	North and South America	Financials	Inactive
98	Indian Oil	India	Asia	Oil & Gas	Inactive
99	Marathon Oil	United States	North and South America	Oil & Gas	Inactive
100	Royal Bank of Scotland	Scotland	North and South America	Financials	Active

Table 5-22: List of Fortune Global 100 Firms for Sample Selection (based on Fortune website (2011) and Industry Qualification Benchmark website (2011))

Data Collection Methodology

Variable	Quote	Codification	Classification
Company information			
Company name	Insert company name from Fortune ranking	No codification	No classification (background information)
Country and region of origin	Insert country and region of origin based on Fortune	According to region	No classification (background information)
Sector	Insert sector based on Industry Qualification Benchmark list	According to Industry Qualification Benchmark	No classification (background information)
Issue prioritization of poverty	Insert themes from headings of sustainability section on website or in CSR	Poverty as prioritization issue/ poverty not as prioritization issue	No classification (background information)
Issue framing of poverty	Use poverty alleviation statements for analysis (see below)	Positive framing on poverty/ neutral framing/ negative framing/ no framing	No classification (background information)
Poverty information: Statement in CSR report or website on…			
…Poverty alleviation in general	Copy information provided	Yes/ no	No classification (background information)
…Engagement in Business-as-Usual	Copy information provided	Yes/ no	No classification (background information)
…Engagement in Classical Corporate Philanthropy	Copy information provided	Yes/ no	Reactive/ inactive
…Engagement in Strategic Corporate Philanthropy	Copy information provided	Yes/ no	Active/ inactive
…Engagement in Narrow Microfinance	Copy information provided	Yes/ no	Reactive/ inactive
…Engagement in Broad Microfinance	Copy information provided	Yes/ no	Active/ inactive
…Engagement in Narrow Bottom of the Pyramid	Copy information provided	Yes/ no	Active/ inactive
…Engagement in Inclusive Bottom of the Pyramid	Copy information provided	Yes/ no	Proactive/ inactive

Variable	Quote	Codification	Classification
...Engagement in Narrow Social Entrepreneurship	Copy information provided	Yes/ no	Active/ inactive
...Engagement in Broad Social Entrepreneurship	Copy information provided	Yes/ no	Proactive/ inactive
...Engagement in Cross-sector Partnerships	Copy information provided	Type of partnering approach (extensive/ moderate/ limited/ none)	Proactive/ active/ reactive/ inactive
...Combination of business strategies	Copy information provided	Yes/ no	No classification (background information)
...Number of strategies combined		6/ 5/ 4/ 3/ 2/ 1/ 0	Proactive (5-6)/ active (3-4)/ reactive (1-2)/ inactive (0)
... Type of strategies combined: CP and MF; CP and BOP; CP and SE; Microfinance and BOP; Microfinance and SE; BOP and SE		Yes/ no (per possible combination)	No classification (background information)
....Combination of business strategies and partnerships	Copy information provided	Yes/ no	No classification (background information)
... Number of strategies combined		4/ 3/ 2/ 1/ 0	Proactive (3-4)/ Active (2)/ Reactive (1)/ Inactive (0)
... Type of strategies combined: CSP and CP; CSP and Microfinance; CSP and BOP; CSP and SE		Yes/ no (per possible combination)	No classification (background information)
Additional useful information and notes	Copy information provided	No coding	No classification (background information)
Poverty Alleviation Portfolio Approach			
Overall Poverty Alleviation Portfolio Approach	None	Extensive/ moderate/ limited/ no Poverty Alleviation Portfolio Approach	Proactive/ active/ reactive/ inactive

Table 5-23: Background Information on the Methodology for Data Collection, Coding and Company Classifications

	A	B	C	D	E	F
1	Rank	Company	Poverty statement quote	Business-as-usual quote	Corporate Philanthropy quote	Microfinance quote
2	1	Wal-Mart Stores	Some of the biggest	Saving people money so	Wal-Mart is choosing to generously	
3	2	Royal Dutch Shell			Shell Foundation is an independent	
4	3	Exxon Mobil	Reducing poverty		Employee volunteerism and giving	In Indonesia, ExxonMobil
5	4	BP	As a member of this body,	We have the potential to be	When BP employees choose to support	We run a range of programmes
6	5	Sinopec Group	We have made positive	Sinopec corp. paid totally R	For the past many years, Sinopec Corp.	
7	6	China National Petroleum	CNPC has been conducting		In-kind donations: the Company actively	
8	7	State Grid	SGCC is an active		We pushed	
9	8	Toyota Motor	In addition, we engage with diverse communities and su		Since 1991, Toyota has contributed over half a billion dollars to philanthrc	
10	9	Japan Post Holdings			through the Japan International	
11	10	Chevron	Poverty is of deep concern		No own foundation mentioned	Microfinancing provides
12	11	Total	We are committed to providing assistance to particularly vulnerable populations in our host region			We have spent more than $15
13	12	ConocoPhillips	ConocoPhillips endorses	Positively impacting	In various locations in the United States,	
14	13	Volkswagen			workforce donations: The "Starthilfe"	
15	14	AXA	For twenty years, all over		As a responsible corporate citizen, AXA st	In addition to AXA's philantrop
16	15	Fannie Mae			Donations to	
17	16	General Electric	The urgency of the		Philanthropic efforts—including donations made by GE Foundation and G	
18	17	ING Group	Being a good corporate	Beyond business as usual	Building on this success, our annual	The term microfinance also ref
19	18	Glencore International		Glencore's global	Community development: the	
20	19	Berkshire Hathaway		Nevertheless, Berkshire's		

Table 5-24: Data Collection Exemplary Excel Sheet Screen Shot

Coding Methodology

Yes for Poverty Statement Provision, if	No for Poverty Statement Provision, if
Explicit provision of a statement (of engagement of firm) with regard to poverty or related terms (e.g. community engagement, marginalization, social exclusion)	No explicit provision of a statement (of engagement of firm) with regard to poverty or related terms (e.g. community engagement, marginalization, social exclusion)

Table 5-25: Codification Methodology for the Provision of a Poverty Statement

Yes for Business-as-Usual, if	No for Business-as-Usual, if
Explicit statement on no engagement in poverty-related activities – operating Business-as-Usual – or	Explicit statement of engagement in poverty-related activities (see above) - operating beyond Business-as-Usual - or
No engagement in poverty-related activities	Engagement in one or more poverty-related activities

Table 5-26: Codification Methodology for the Engagement in Business-as-Usual

Yes for strategic CP, if	Yes for classical CP, if	No, if
Provision of monetary and non-monetary donations	Provision of solely monetary donations	No statement on engagement in (classical or strategic) corporate philanthropy
Existence of own foundation	No existence of own foundation	
Employee volunteering and/ or employee giving	No employee volunteering or employee giving	
Activities highly related to poverty alleviation	Activities solely partially related to poverty alleviation	
Broad number of philanthropic activities	Limited number of philanthropic activities	
Philanthropic activities are related to/ part of core business	Philanthropic activities are unrelated to core business	
Engagement in philanthropic partnerships	No engagement in philanthropic partnerships	

Table 5-27: Codification Methodology for the Engagement in Corporate Philanthropy

Yes for broad Microfinance, if	Yes for narrow Microfinance, if	No, if
Provision of additional microfinance services	Provision of solely microcredit services	No statement on engagement in (narrow or broad) microfinance
Provision of training or additional support services	No provision of training or additional support services	
Broad number of microfinance activities	Limited number of microfinance activities	
Microfinance activities are related to/ part of core business	Microfinance activities are unrelated to core business	
Engagement in microfinance partnerships	No engagement in microfinance partnerships	

Table 5-28: Codification Methodology for the Engagement in Microfinance

Yes for inclusive BOP, if	Yes for narrow BOP, if	No, if
Engagement of poor as well as producers, employees or entrepreneurs	Engagement of poor as customer	No statement on engagement in (narrow or inclusive) bottom of the pyramid
Limited number of BOP activities	Broad number of BOP activities	
BOP activities are unrelated to core business	BOP activities are related to/ part of core business	
No engagement in BOP partnerships	Engagement in BOP partnerships	

Table 5-29: Codification Methodology for the Engagement in Bottom of the Pyramid

Yes for broad SE, if	Yes for narrow SE, if	No, if
Engagement into own SE activities	Sole support of external social businesses	No statement on engagement in (narrow or broad) social entrepreneurship
Activities highly related to poverty alleviation	Activities solely partially related to poverty alleviation	
Broad number of SE activities	Limited number of SE activities	
SE activities are related to/ part of core business	SE activities are unrelated to core business	
Engagement in SE partnerships	No engagement in SE partnerships	

Table 5-30: Codification Methodology for the Engagement in Social Entrepreneurship

Extensive, if	Moderate, if	Limited, if	No, if
High in number (4 or more projects)	Moderate in number (2-3 projects)	Few in number (1 project)	No statement on engagement in cross-sector partnerships
High amount of partners (3 or more)	Moderate amount of partners (2)	Low amount of partners (1)	
High in duration (more than 2 years)	Moderate in duration (1-2 years)	Short in duration (less than 1 year)	
Extensive in scope (3 or more countries; 3 or more themes)	Moderate in scope (2-3 countries; 2-3 themes)	Narrow in scope (1 country; 1 theme)	
Extensive form of engagement (financial and non-financial contribution)	Moderate form of engagement (primarily financial contribution)	Limited form of engagement (solely financial contribution)	

Table 5-31: Codification Methodology for the Engagement in Cross-Sector Partnerships

Yes, if	No, if	Number of strategies combined	Types of strategies combined
One or more combination of strategies	No combination of strategies	State number of combined strategies (0, 1, 2, 3, 4, 5, 6)	State type of combined strategies (CP and Microfinance, CP and BOP, CP and SE, Microfinance and BOP, Microfinance and SE, BOP and SE)

Table 5-32: Codification Methodology for the Combination of Business Strategies

Yes, if	No, if	Number of approaches combined	Types of approaches combined
One or more combination of strategies with CSPs	No combination of strategies with cross-sector partnerships	State number of combined approaches (0, 1, 2, 3, 4)	State type of approaches combined (CSP and CP, CSP and Microfinance, CSP and BOP, CSP and SE)

Table 5-33: Codification Methodology for the Combination of Cross-Sector Partnerships and Business Strategies

Issue prioritization of poverty, if	No issue prioritization of poverty, if
Poverty listed as CSR focus theme on website	Poverty not listed as CSR focus theme on website

Table 5-34: Codification Methodology for the Issue Prioritization of Poverty

Issue framing in positive terms	Issue framing in neutral terms	Issue framing in negative terms	No issue framing on poverty
Poverty described positively as opportunity; focus on contributions and solutions	Poverty described neutrally or in both negative and positive terms	Poverty described in negative terms as threat; focus on negative effects and risks	No statement on poverty made

Table 5-35: Codification Methodology for the Issue Framing of Poverty

Company Classification Methodology

Variable	Codification	Classification
Sub-classifications		
Engagement in Corporate Philanthropy	Yes for Strategic CP	Active (2)
	Yes for Classical CP	Reactive (1)
	No for Strategic or Classical CP	Inactive (0)
Engagement in Microfinance	Yes for Broad Microfinance	Active (2)
	Yes for Narrow Microfinance	Reactive (1)
	No for Broad or Narrow Microfinance	Inactive (0)
Engagement in Bottom of the Pyramid	Yes for Inclusive BOP	Proactive (3)
	Yes for Narrow BOP	Active (2)
	No for Inclusive or Narrow BOP	Inactive (0)
Engagement in Social Entrepreneurship	Yes for Broad SE	Proactive (3)
	Yes for Narrow SE	Active (2)
	No for Broad or Narrow SE	Inactive (0)
Engagement in Cross-sector Partnerships	Yes for extensive CSPs	Proactive (3)
	Yes for moderate CSPs	Active (2)
	Yes for limited CSPs	Reactive (1)
	No for CSPs	Inactive (0)
Number of combined of business strategies	5-6	Proactive (3)
	3-4	Active (2)
	1-2	Reactive (1)
	0	Inactive (0)
Number of combined business strategies and cross-sector partnerships	3-4	Proactive (3)
	2	Active (2)
	1	Reactive (1)
	0	Inactive (0)
Overall Classification		
Poverty alleviation portfolio approach (sum of the 7 prior sub-classification)	Sum between 15 and 19	***Proactive poverty alleviation portfolio approach***
	Sum between 10 and 14	***Active poverty alleviation portfolio approach***
	Sum between 5 and 9	***Reactive poverty alleviation portfolio approach***
	Sum between 0 and 4	***Inactive poverty alleviation portfolio approach***

Table 5-36: Background Information on the Methodology for the Company Classification

	A	B	C	D	E
1	Rank	Company	Corporate Philanthropy quote	Score	Classification
2	1	Wal-Mart Stores	" Wal-Mart is choosing to generously support a	2	Active
3	2	Royal Dutch Shell	Shell Foundation is an independent charity that	2	Active
4	3	Exxon Mobil	Employee volunteerism and giving Employees and	1	Reactive
5	4	BP	When BP employees choose to support charitable	1	Reactive
6	5	Sinopec Group	For the past many years, Sinopec Corp. has	2	Active
7	6	China National Petroleum	In-kind donations: the Company actively	2	Active
8	7	State Grid	We pushed	2	Active
9	8	Toyota Motor	Since 1991, Toyota has contributed over half a billion	1	Reactive
10	9	Japan Post Holdings	through the Japan International Cooperation	1	Reactive
11	10	Chevron		0	Inactive
12	11	Total	Created in 1992, the Total Foundation focuses on	2	Active
13	12	ConocoPhillips	In various locations in the United States,	1	Reactive
14	13	Volkswagen	workforce donations: The "Starthilfe" (Getting	1	Reactive
15	14	AXA	As a responsible corporate citizen, AXA strives to pla	2	Active
16	15	Fannie Mae	Donations to	1	Reactive
17	16	General Electric	Philanthropic efforts—including donations made by	1	Reactive
18	17	ING Group	Building on this success, our annual Global	2	Active
19	18	Glencore International	Community development: the programmes	1	Reactive
20	19	Berkshire Hathaway		0	Inactive
21	20	General Motors	donations to community development and poverty a	1	Reactive
22	21	Bank of America Corporation	corporate volunteering; donations are provided to	1	Reactive
23	22	Samsung Electronics	engagement in corporate volunteering; donations	1	Reactive
24	23	ENI	In Brazil, where the situation is	2	Active
25	24	Daimler	Volunteering and other activities for the common	2	Active
26	25	Ford Motor	Founded as a not-for-profit organization in 1949,	2	Active

Table 5-37: Company Classification Exemplary Excel Screen Shot

Rank	Company	CP Score	MF Score	BOP Score	SE Score	CSP Score	BS Combi Score	BA Combi Score	Total Score	Classification
1	Wal-Mart Stores	2	0	0	0	1	0	1	4	Inactive
2	Royal Dutch Shell	2	0	0	0	1	0	1	4	Inactive
3	Exxon Mobil	1	2	3	0	3	0	1	10	Active
4	BP	1	2	3	0	2	0	0	8	Reactive
5	Sinopec Group	2	0	0	0	1	0	0	3	Inactive
6	China National Petroleum	2	0	0	0	0	0	0	2	Inactive
7	State Grid	2	0	0	0	1	0	0	3	Inactive
8	Toyota Motor	1	0	0	0	1	0	0	2	Inactive
9	Japan Post Holdings	1	0	0	0	0	0	0	1	Inactive
10	Chevron	0	2	0	0	3	0	1	6	Reactive
11	Total	2	1	0	0	1	1	1	6	Reactive
12	ConocoPhillips	1	0	0	0	0	0	0	1	Inactive
13	Volkswagen	1	0	0	0	0	0	0	1	Inactive
14	AXA	2	1	2	0	2	0	1	8	Reactive
15	Fannie Mae	1	0	0	0	0	0	0	1	Inactive
16	General Electric	1	0	3	2	1	2	1	10	Active
17	ING Group	2	2	0	0	2	0	0	6	Reactive
18	Glencore International	1	0	0	0	0	0	0	1	Inactive
19	Berkshire Hathaway	0	0	0	0	0	0	0	0	Inactive
20	General Motors	1	0	0	0	1	0	0	2	Inactive
21	Bank of America Corporation	1	1	0	0	3	0	1	6	Reactive
22	Samsung Electronics	1	0	0	0	3	0	0	4	Inactive
23	ENI	2	2	0	0	3	0	1	8	Reactive
24	Daimler	2	2	0	0	1	0	1	6	Reactive
25	Ford Motor	2	0	0	0	1	0	0	3	Inactive
26	BNP Paribas	2	1	0	2	3	0	2	10	Active

Table 5-38: Overall Company Classification Excel Overview

Appendix G: Practical Part – Results and Discussion

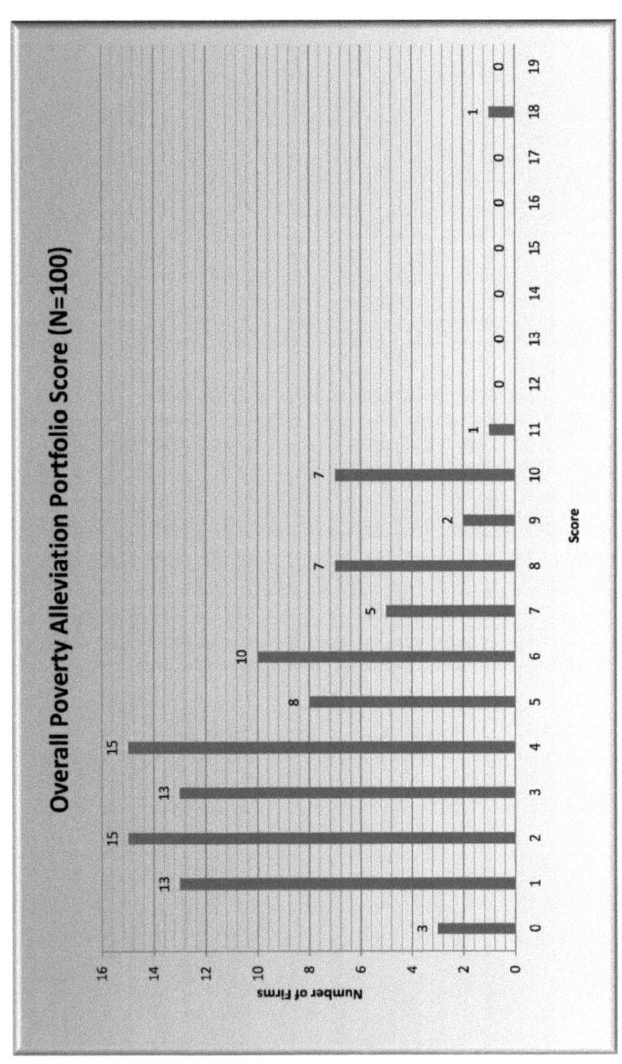

Figure 5-2: Overall Poverty Alleviation Portfolio Score

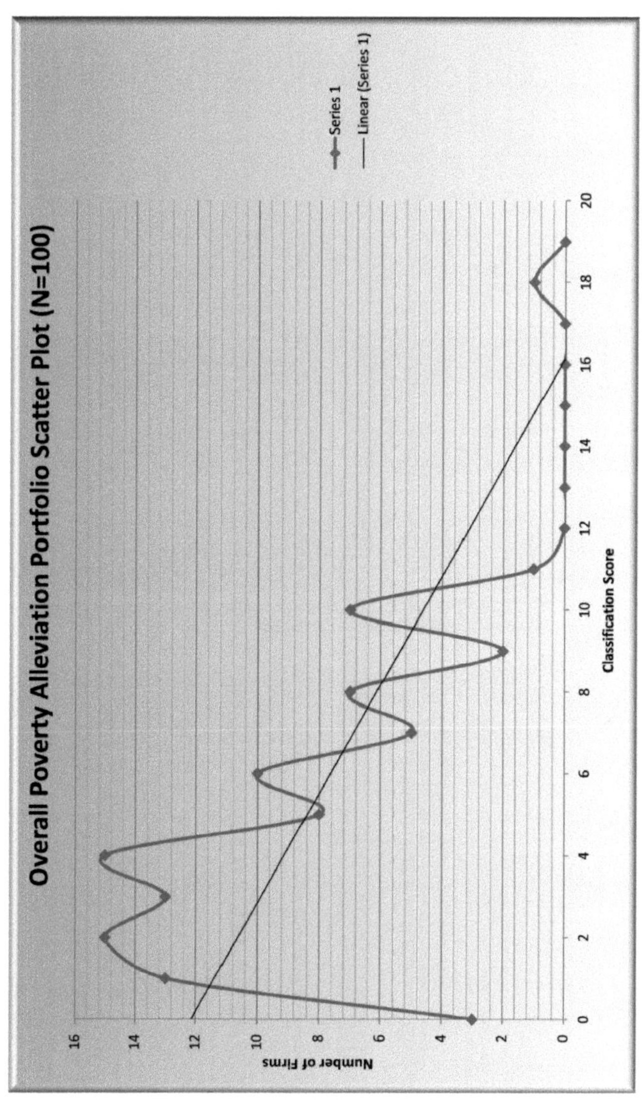

Figure 5-3: Overall Poverty Alleviation Portfolio Scatter Plot

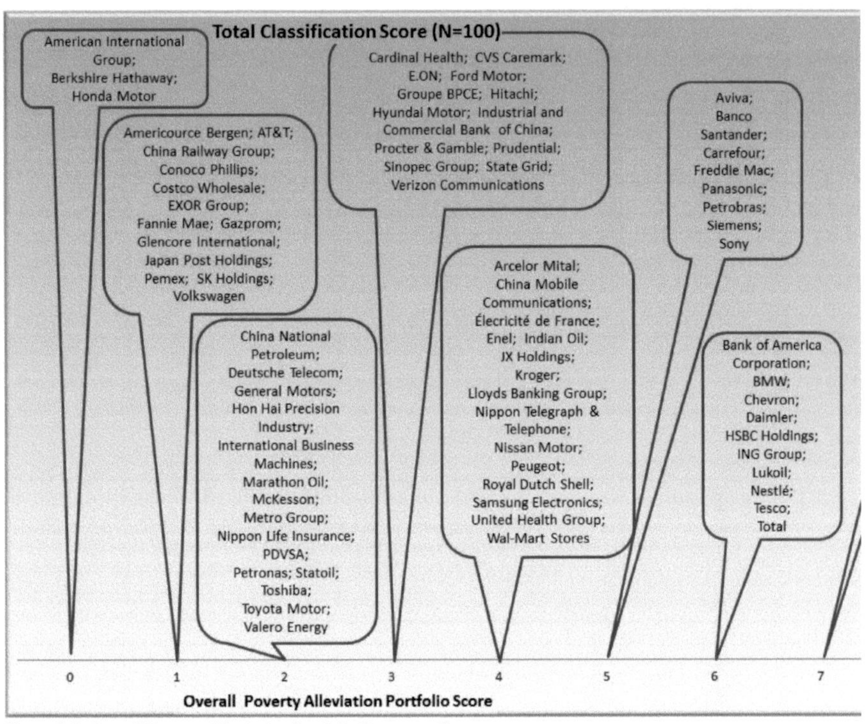

Figure 5-4: Overall Poverty Alleviation Portfolio Score per Company

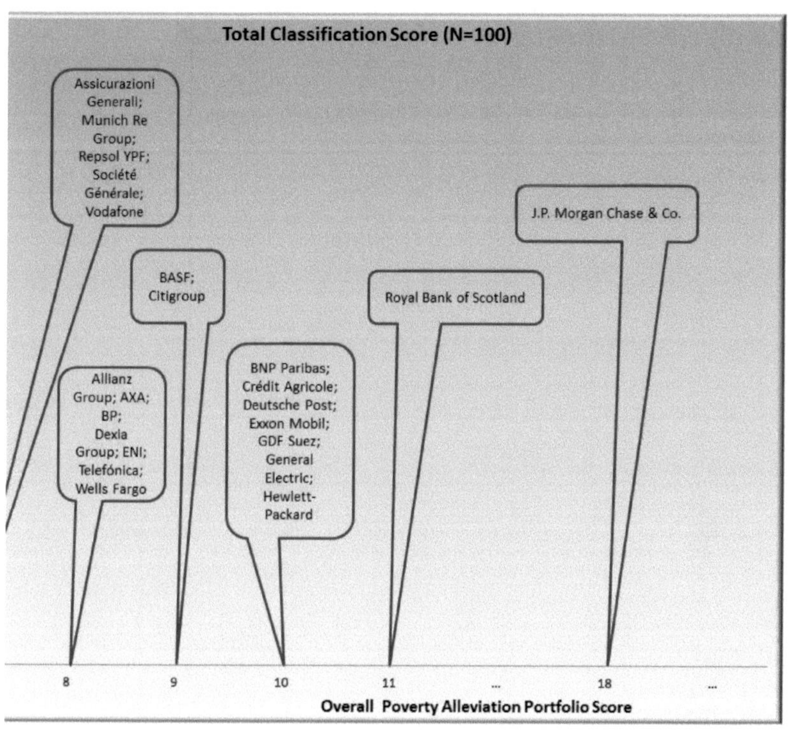
Total Classification Score (N=100)
Assicurazioni Generali; Munich Re Group; Repsol YPF; Société Générale; Vodafone
BASF; Citigroup
Allianz Group; AXA; BP; Dexia Group; ENI; Telefónica; Wells Fargo
BNP Paribas; Crédit Agricole; Deutsche Post; Exxon Mobil; GDF Suez; General Electric; Hewlett-Packard
Royal Bank of Scotland
J.P. Morgan Chase & Co.
8
9
10
11
...
18
...
Overall Poverty Alleviation Portfolio Score

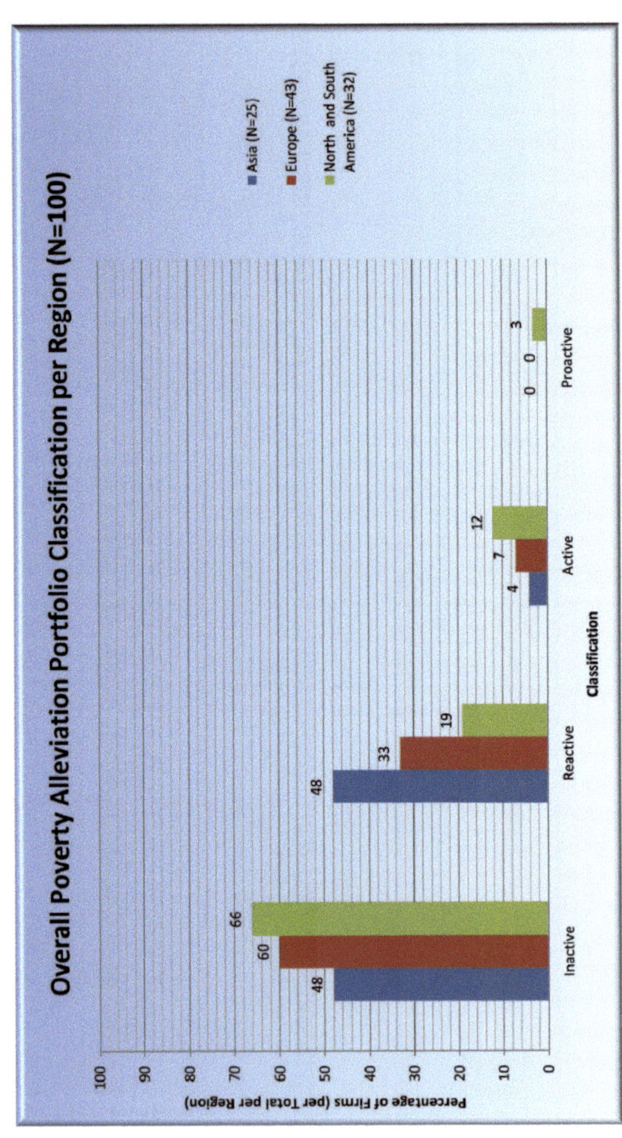

Figure 5-5: Overall Poverty Alleviation Portfolio Classification per Region

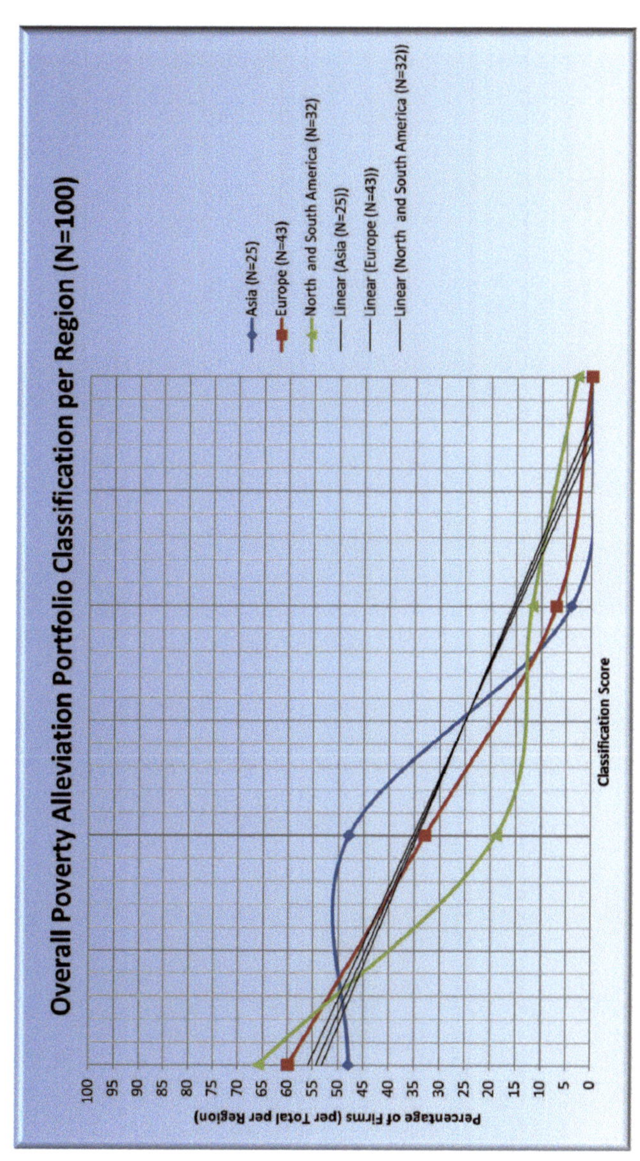

Figure 5-6: Overall Poverty Alleviation Portfolio Scatter Plot per Region

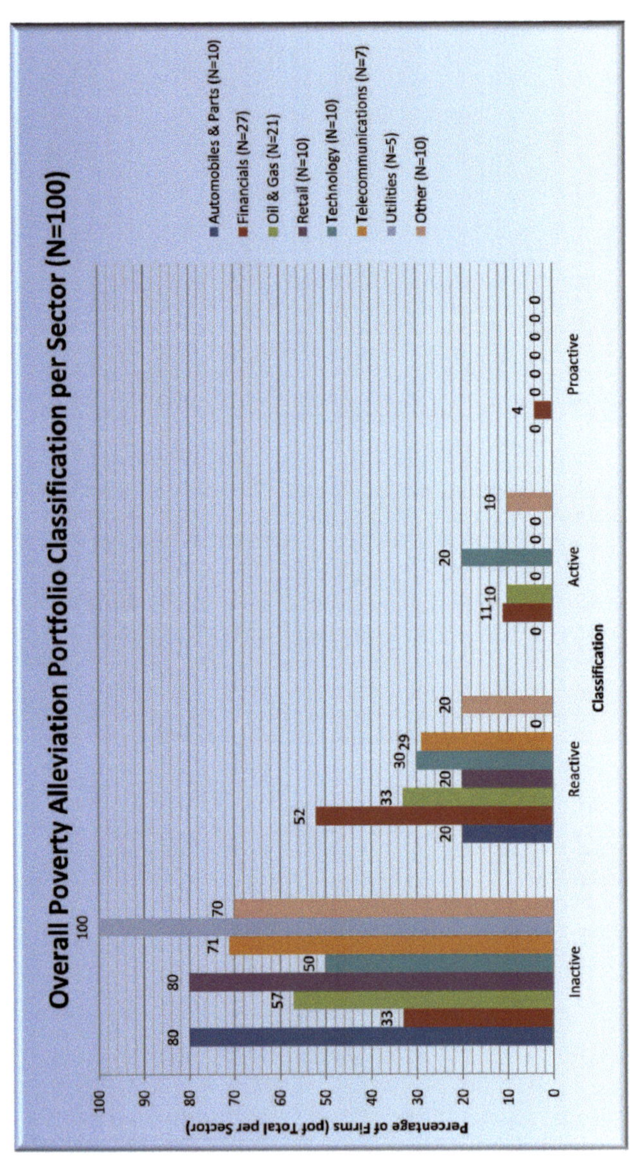

Figure 5-7: Overall Poverty Alleviation Portfolio Classification per Sector

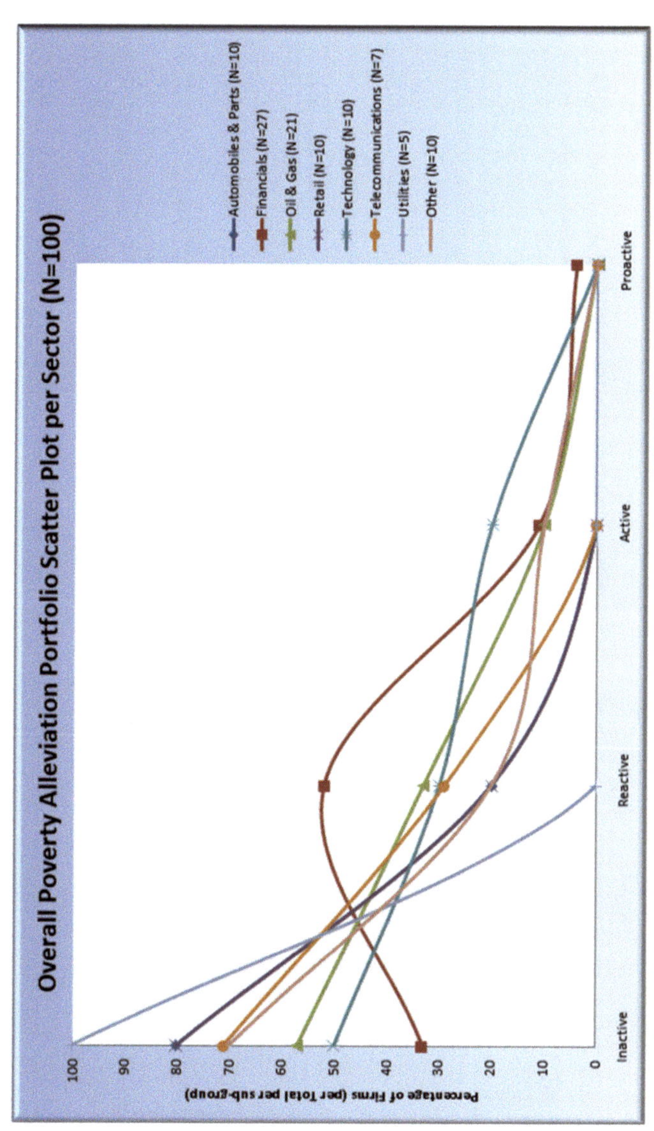

Figure 5-8: Overall Poverty Alleviation Portfolio Scatter Plot per Sector

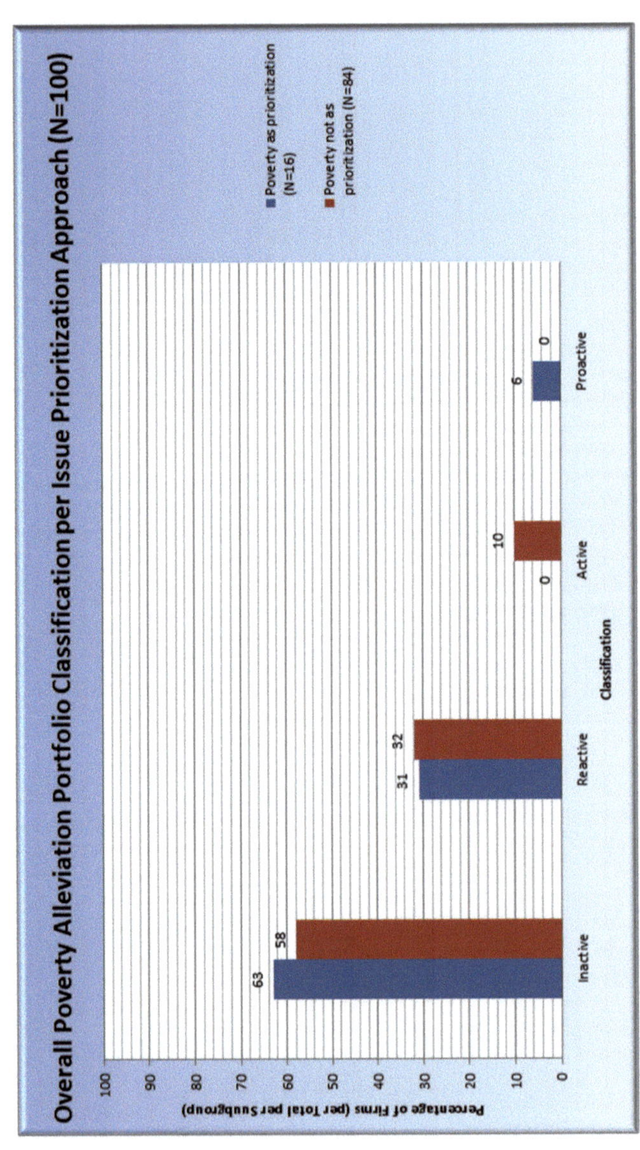

Figure 5-9: Overall Poverty Alleviation Portfolio Classification per Issue Prioritization Approach

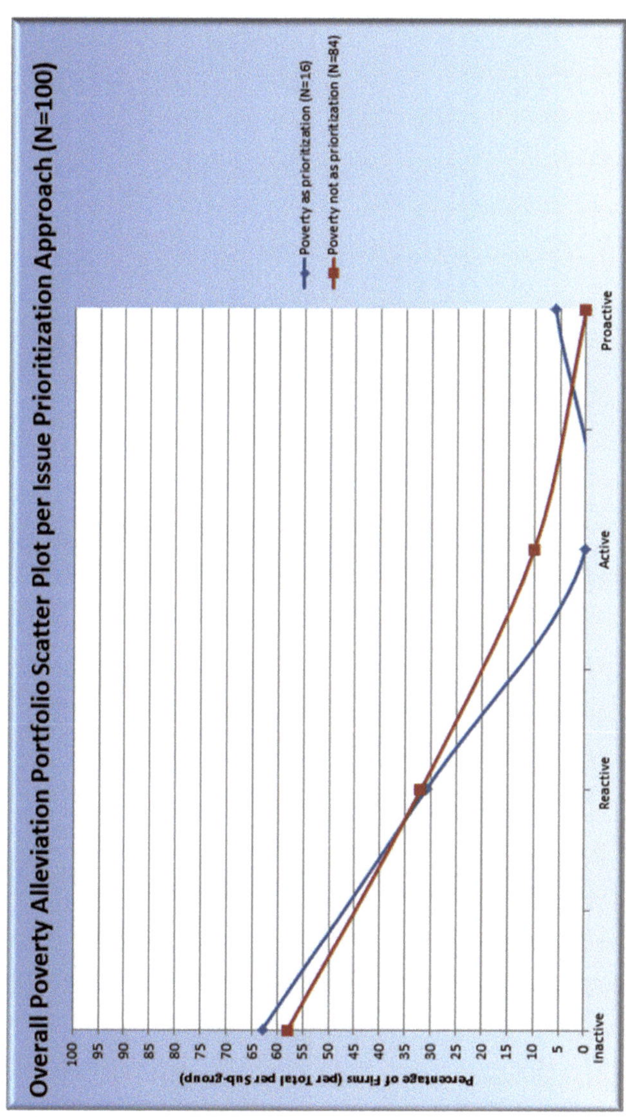

Figure 5-10: Overall Poverty Alleviation Portfolio Scatter Plot per Issue Prioritization Approach

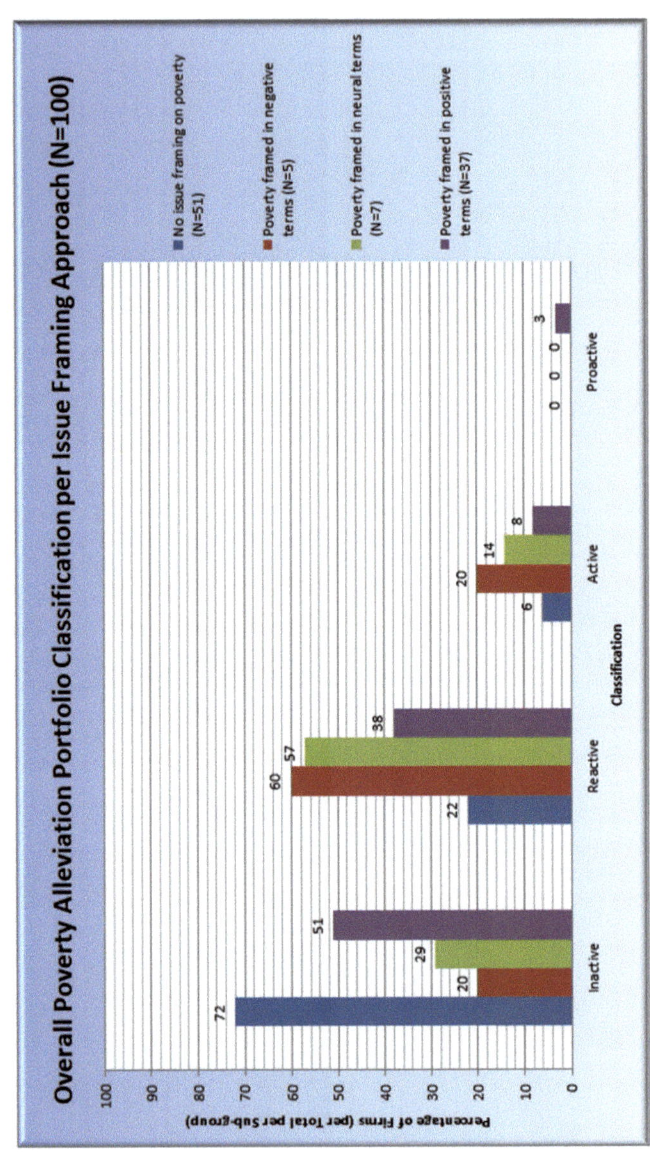

Figure 5-11: Overall Poverty Alleviation Portfolio Classification per Issue Framing Approach

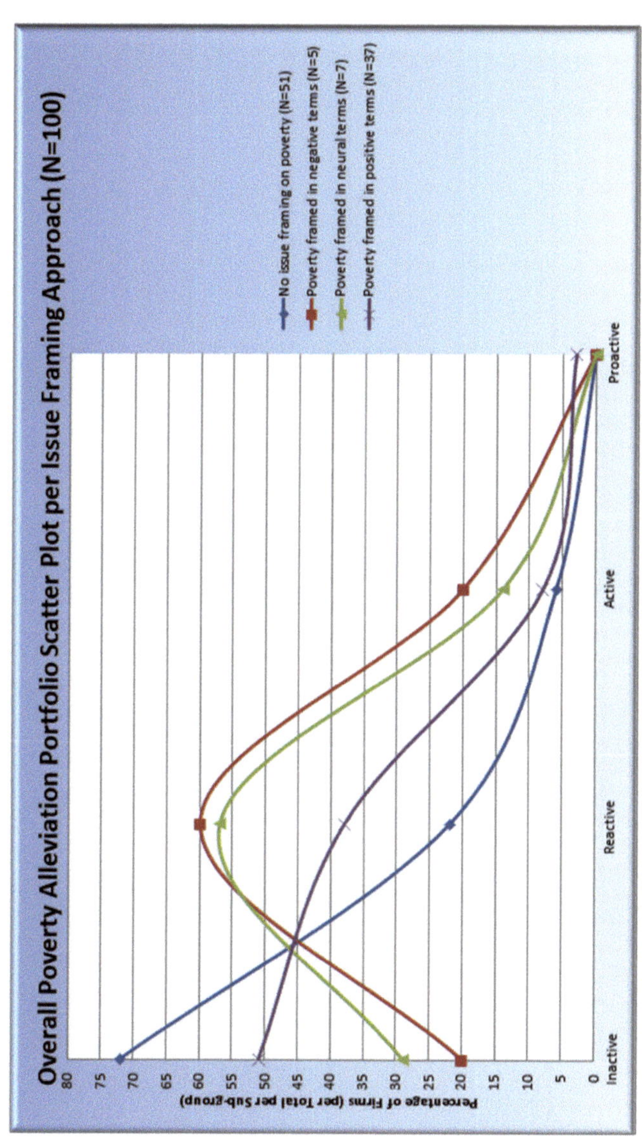

Figure 5-12: Overall Poverty Alleviation Portfolio Scatter Plot per Issue Framing Approach

Bibliography

Scientific Articles, Books and Other Publications

Andrews, R. & Entwistle, T. (2010). Does Cross-Sectorial Partnership Deliver? An Empirical Exploration of Public Service Effectiveness, Efficiency and Equity. *Journal of Public Administration Research and Theory,* 20, pp. 679-701.

Arya, B. & Salk, J. E. (2006). Cross-sector alliance learning and effectiveness of voluntary codes of corporate social responsibility. *Business Ethics Quarterly,* 16 (2), pp. 211-234.

Austin, J, Stevenson, H. & Wei-Skillern, J. (2006). Social and Commercial Entrepreneurship: Same, Different or Both? *Entrepreneurship Theory and Practice,* pp. 1-22.

Bais, K. (2008). The Base of the Pyramid as a Development Strategy. *Oxfam Novib*, 3, pp. 1-18.

Barney, I. (2003). Business, community development and sustainable livelihood approaches. *Community Development Journal,* 38 (3), pp. 255-265.

Baron, R. M. & Kenny, D. A. (1986). The moderator-mediator variable distinction in social psychological research: Conceptual, strategic and statistical considerations. *Journal of Personality and Social Psychology,* 51, pp. 1173-1182.

Belal, A. R. & Cooper, S. (2007). *The Absence of Corporate Social Responsibility in Bangladesh.* Birmingham: Aston Business School.

Bhowmick, S. (2011). Social Cause Venturing as a Distinct Domain. *Journal of Social Entrepreneurship,* 2 (1), pp. 99-111.

Bishop, M. (2007). Fighting Global Poverty: Who'll Be Relevant in 2020? *The Economist,* 4, pp. 1-10.

Blowfield, M. & Frynas, J. G. (2005). Setting new agendas: critical perspectives on Corporate Social Responsibility in the developing world. *International Affairs,* 81 (3), pp. 499-513.

Brinkerhoff, D. W. & Brinkerhoff, J. M. (2011). Public-Private Partnerships: Perspectives on Purposes, Publicness and Good Governance. *Public Administration and Development,* 31, pp. 2-14.

Brinkerhof, J. M. (2002). Assessing and improving partnership relations and outcomes: A proposed framework. *Evaluation and Program Planning,* 25, pp. 215-231.

Bryman, A. & Bell, E. (2007). *Business Research Methods.* 2nd edition. New York: Oxford University Press.

Campbell, H. (2005). Business economic impacts: The new frontier for corporate accountability. *Development in Practice,* 15 (3 & 4), pp. 413-421.

Chakraborty, S. K., Kurien, V., Singh, J., Atheya, M., Maira, A., Aga, A. & Gupta, A. K. (2004). Management Paradigms Beyond Profit Maximization. *Vikalpa,* 29 (3), pp. 97-117.

Chapple, W. & Moon, J. (2005). Corporate Social Responsibility (CSR) in Asia – A Seven-Country Study of Web Site Reporting. *Business and Society,* 44, pp. 415-441.

Christensen, L. (2008). Alleviating Poverty Using Microfranchising Models: Case Studies and a Critique. In Wankel, C. (Eds.). *Alleviating Poverty through Business Strategy: Global Case Studies in Social Entrepreneurship.* New York: Palgrave Macmillan.

Cochran, P. L. & Wood, R. A. (1984). Corporate Social Responsibility and Financial Performance. *Academy of Management Journal,* 27 (1), pp. 42-56.

Council of the European Union (2004). *Joint Report by the Commission and the Council on Social Inclusion.* Brussels: EU.

Coxon, E. & Munce, K. (2008). The global education agenda and the delivery of aid to Pacific education. *Comparative Education,* 44 (2), pp. 147-165.

Curtis, D. (2006). Mind sets and methods. Poverty strategies and the awkward potential of the enabling state. *International Journal of Public Sector Management,* 19 (2), pp. 150-164.

Cycyota, C. S. & Volkland, W. (2007). Sustainable Workforce Models: Lessons from India on Training and Development of Unskilled Labor. In Stoner, J. A. F. & Wankel, C. (2007). *Innovative Approaches to Reducing Global Poverty.* Charlotte: Information Age Publishing.

da Rosa, A., Barendse, L. & Kozarjan, G. (2011). *A portfolio approach to cross-sector partnerships – Fact Sheets from the strategic alliance literature.* Rotterdam: The Partnerships Resource Center.

da Rosa, A. & van Tulder, R. (2011). *Measuring the Impact of Partnerships on Poverty Reduction.* Rotterdam: The Partnership Resource Center.

Dawkins, J. & Lewis, S. (2003). CSR in Stakeholder Expectations: And Their Implication for Company Strategy. *Journal of Business Ethics,* 44, pp. 185-193.

Dees, J. G. (2001). The Meaning of Social Entrepreneurship. *Drew Charter School,* pp. 1-16.

Devarajan, S. & Reinikka, R. (2004). Making services work for poor people. *Journal of African Economies*, 13 (suppl. 1), pp. 142-166.

Dolan, C. S. & Scott, L. (2009). Lipstick evangelium: Avon trading circles and gender empowerment in South Africa. *Gender and Development,* 17 (2), pp. 203-218.

Dollery, B. & Wallis, J. (2001). Social Service Delivery and the Voluntary Sector in Contemporary Australia. *Working Paper Series in Economics,* pp.1-10.

Douglas, A. (2008). Beyond aid: the future of international development. *Public Policy Research,* 2, pp. 48-51.

Easterly, L. & Miesing, P. (2007). Social Venture Business Strategy for Reducing Poverty. In Stoner, J. A. F. & Wankel, C. (2007). *Innovative Approaches to Reducing Global Poverty.* Charlotte: Information Age Publishing.

Eder, M. & Öz, O. (2008). Scrutinizing the Link between Poverty and Business Strategy: What Can We Learn from the Case of Shuttle Traders in Laleli, Istanbul? In Wankel, C. (Eds.). *Alleviating Poverty through Business Strategy: Global Case Studies in Social Entrepreneurship.* New York: Palgrave Macmillan.

Farias, C. & Farias, G. (2008). Cycles of poverty and consumption: The sustainability dilemma. *Competitive Review,* 20 (3), pp. 248-257.

Fortanier, F. & Kolk, A. (2007). On the Economic Dimensions of Corporate Social Responsibility: Exploring Fortune Global 250 Reports. *Business and Society,* 46 (4), pp. 457-478.

Franceys, R. & Weitz, A. (2003). Public-Private Community Partnerships in Infrastructure for the Poor. *Journal of International Development,* 15, pp. 1083-1098.

Friedman, M. (1970). The social responsibility of business is to increase profits. *New York Times Magazine,* 33, pp. 122-126.

Frynas, J. G. (2008). Corporate Social Responsibility and International Development: Critical Assessment. *Corporate Governance,* 16 (4), pp. 274-281.

Fukuda-Parr, S. (2004). The Millennium Development Goals: The Pledge of World Leaders to End Poverty Will Not Be Met with Business as Usual. *Journal of International Development,* 16, pp. 925-932.

Gardetti, M. A. (2005). A Base-of-the-Pyramid Approach in Argentina: Preliminary Findings from a BOP Learning Lab. *Greener Management International,* 51, pp. 65-77.

Gifford, B., Krestler, A. & Anand, S. (2010). Building local legitimacy into corporate social responsibility: Gold mining firms in developing nations. *Journal of World Business,* 45, pp. 304-311.

Ginsberg, B. (2002). Even International Summits Must Be Localized: A Youth Perspective on the UN World Summit on Sustainable Development. *Local Environment,* 7 (4), pp. 471-476.

Gold, L. (2004). The 'Economy of Communication': A case study of business and civil society in partnership for change. *Development in Practice,* 14 (5), pp. 633-644.

Goldsmith, A. A. (2011). Profits and Alms: Cross-Sector Partnerships for Global Poverty Reduction. *Public Administration and Development,* 31, pp. 15-24.

Gore, C. (2010). The MDG Paradigm, Productive Capacities and the Future of Poverty Reduction. *IDS Bulletin,* 41 (1), pp. 70-79.

Hahn, R. (2009). The Ethical Rational of Business for the Poor – Integrating the Concepts Bottom of the Pyramid, Sustainable Development, and Corporate Citizenship. *Journal of Business Ethics,* 84, pp. 313-324.

Helsin, P. A. & Ochoa, J. D. (2008). Understanding and developing strategic corporate social responsibility. *Organizational Dynamics,* 37 (2), pp. 125-144.

Hermann, K. (2003). *Linking small with big – Measuring the impact of private sector involvement in poverty reduction and local economic development.* London: DFID.

Heslam, P. S. (2007). Reducing Poverty Through Successful Business: The Role of Social Capital. In Stoner, J. A. F. & Wankel, C. (2007). *Innovative Approaches to Reducing Global Poverty.* Chalotte: Information Age Publishing.

Hossain, F. & Knight, T. (2008). Can micro-credit improve the livelihood of the poor and disadvantaged? Empirical observations from Bangladesh. *International Development Planning Review,* 30 (2), pp. 155-176.

Hossain, I. (2000). Micro-Credit and Good Governance: Models of Poverty Alleviation. *Southeast Asian Journal of Social Science,* 28 (1), pp. 185-208.

Hotel, C. (2005). *The Multidimensionality of Poverty: An Institutionalist Perspective. Conference Paper in International Conference 'The many dimensions of poverty'.* Brasilia: International Poverty Center of the United Nations Development Programme.

Humphrey, J. (2006). Prospects and Challenges for Growth and Poverty Reduction in Asia. *Development Policy Review,* 24 (1), pp. 29-49.

ICCO (2008). *The Base of the Pyramid as a development strategy.* Utrecht: Icco.

Ite, U. E. (2007). Changing Times and Strategies: Shell's Contribution to Sustainable Community Development in the Niger Delta, Nigeria. *Sustainable Development,* 15, pp. 1-14.

Jenkins, R. (2005). Globalization, Corporate Social Responsibility and Poverty. *International Affairs,* 81 (3), pp. 525-540.

Johnson, S. (2009). Microfinance is dead! Long live microfinance! Critical reflections on two decades of microfinance policy and practice. *Enterprise Development and Microfinance,* 20 (4), pp. 291-303.

Kanji, N. (2004). Corporate responsibility and women's employment: The case of cashew nuts. *Gender and Development,* 12 (2), pp. 82-87.

Karlan, D. & Valdivia, M. (2011). Teaching Entrepreneurship: Impact of Business Training on Microfinance Clients and Institutions. *The Review of Economics and Statistics,* 93 (2), pp. 510-527.

Karnani, A. (2007a). *Employment, not microcredit, is the solution.* Michigan: University of Michigan.

Karnani, A. (2007b). Misfortune at the bottom of the pyramid. *Greener Management International,* 51, pp. 99-110.

Karnani, A. (2010). Failure of the libertarian approach to reducing poverty. *Asian Business and Management,* 9 (1), pp. 5-21.

Kelley, S., Werhane, P. H. & Hartmann, L. P. (2008). The End of Foreign Aid as We Know It: The Profitable Alleviation of Poverty in a Globalized Economy. In Wankel, C. (Eds.). *Alleviating Poverty through Business Strategy: Global Case Studies in Social Entrepreneurship.* New York: Palgrave Macmillan.

Kolk, A. & van Tulder, R. (2005). Poverty Alleviation as a Business Strategy? Evaluating Commitments of Frontrunner Multinational Corporations. *World Development,* 34 (5), pp. 789-801.

Kolk, A. & van Tulder, R. (2010). International business, corporate social responsibility and sustainable development. *International Business Review,* 19, pp. 119-125.

Kolk, A, van Tulder, R. & Kostwinder, E. (2008). Business and partnerships for development. *European Management Journal,* 26, pp. 262-273.

Lapeyre, R. (2011). The Grootberg lodge partnership in Namibia: Towards poverty alleviation and empowerment for long-term sustainability? *Current Issues in Tourism,* 14 (3), pp. 221-234.

Leisinger, K. M. (2007). Corporate Philanthropy. The "Top of the Pyramid". *Business and Society Review,* 112 (3), pp. 315-342.

Lodge, G. & Wilson, C. (2006). *A corporate solution to global poverty: How multinationals can help the poor and invigorate their own legitimacy.* Princeton: Princeton University Press.

London, T. (2007). A Base-of-the-Pyramid Perspective on Poverty Alleviation. *The William Davidson Institute,* pp. 1-46.

Lyon, F. & Bertotti, M. (2007). Measuring the Contribution of Small Firms to Reducing Poverty and Increasing Social Inclusion in the United Kingdom. In Stoner, J. A. F. & Wankel, C. (2007). *Innovative Approaches to Reducing Global Poverty.* Charlotte: Information Age Publishing.

Mahantly, S., Yasmi, Y., Guernier, J., Ukkerman, R. & Nass, L. (2009). Relationships, learning and trust: lessons from the SNV-RECOFTC partnership. *Development in Practice,* 19 (7), pp. 859-872.

Mair, J. & Marti, I. (2004). Social Entrepreneurship Research: A Source of Explaination, Prediction and Delight. *IESE Business School,* pp. 1-19.

Mair, J. & Schoen, O. (2005). Social entrepreneurial business models: An exploratory story. *University of Navara,* pp. 1-20.

Mair, J. & Schoen, O. (2007). Successful social entrepreneurial business models in the context of developing economies. *International Journal of Emerging Markets,* 2 (1), pp. 54-68.

Makita, R. (2006). Exploring partnership enterprise for the rural poor through an experimental poultry program in Bangladesh. *Journal of Developmental Entrepreneurship,* 12 (2), pp. 217-237.

Makita, R. (2011). A confluence of Fair Trade and organic agriculture in southern India. *Development in Practice,* 21, (2), pp. 205-217.

Martin, R. & Osberg, S. (2007). Social Entrepreneurship: The Case for Definition. *Stanford Social Innovation Review,* 5 (2), pp. 28-39.

McCord, A *(2004). Public works as a component of social protection in South Africa.Cape Town: Basic Income Grant Coalition.*

McFalls, R. (2007). *Testing the Limits of Inclusive Capitalism: A Case Study of the South Africa HP i-Community.* Stellenbosch: University of Stellenbosch.

McIntyre, S., Bruno, A. & Guerra, P. (2008). In Search of Sustainable Social Mission Ventures to Alleviate Poverty. In Wankel, C. (Eds.). *Alleviating Poverty through Business Strategy: Global Case Studies in Social Entrepreneurship.* Charlotte: Information Age Publishing.

McKague, K., Snelgrove, A. Prada, M., Medalye, J., Stein, G., Alam, R. & Greenspoon, J. (2004). *The Private Sector & Development: Scan of Activities & Research.* York: International Development Research Center.

McMullen, J. S. (2011). Delineating the Domain of Development Entrepreneurship: A Market-Based Approach to Facilitate Inclusive Economic Growth. *Entrepreneurship Theory and Practice,* pp. 185-213.

McWilliams, A. & Siegel, D. (2001). Corporate Social Responsibility: A Theory of the Firm Perspective. *Academy of Management Review,* 28 (1), pp. 117-127.

Merino, A. & Valor, C. (2011). The potential of Corporate Social Responsibility to eradicate poverty: An ongoing debate. *Development in Practice,* 21 (2), pp. 157-167.

Mert, A. (2009). Partnerships for sustainable development as discursive practice: Shifts in discourses of environment and democracy. *Forest Policy and Economics,* 11, pp. 326-339.

Mittal, V. & Ross, Jr. W. T. (1998). The Impact of Positive and Negative Affect and Issue Framing on Issue Interpretation and Risk Taking. *Organizational Behavior and Human Decision Processes,* 76 (3), pp. 298-324.

Montgomery, R. H., Palma, A. & Hoagland-Grey, H. (2008). Community Investment Programs in Developing Infrastructure Projects. *Journal of Infrastructure Systems,* 14 (3), pp. 241-254.

Moran, M. (2008). New Developments in Philanthropy: How Private Foundations Are Changing International Development. *International Trade Forum*, pp. 26-27.

Mosley, P. & Hulme, D. (1998). Microenterprise Finance: Is There a Conflict Between Growth and Poverty Alleviation? *World Development,* 26 (5), pp. 783-790.

Newell, P. & Frynas, J. G. (2007). Beyond CSR? Business, poverty and social justice: An introduction. *Third World Quarterly,* 28 (4), pp. 669-681.

Norton, S. W. (2002). Economic Growth and Poverty: In Search of Trickle-Down. *Cato Journal,* 22 (2), pp. 263-275.

Nunnenkamp, P. (2004). *To what extent can foreign direct investment help achieve international development goals.* Oxford: Blackwell Publishing.

O'Dwyer, B. (2002). Conceptions of Corporate Social Responsibility: The Nature of Managerial Capture. *Accounting, Auditing and Accountability Journal,* 16 (4), pp. 523-557.

Oetzel, J. & Doh, J. P. (2009). MNEs and development: A review and reconceptualization. *Journal of World Business,* 44, pp. 108-120.

Oketch, M. O. (2004). The corporate stake in social cohesion. *Corporate Governance,* 4 (3), pp. 5-19.

Olsen, M. & Boxenbaum, E. (2009). Bottom-of-the-Pyramid: Organizational Barriers to Implementation. *California Management Review,* 51 (4), pp. 100-125.

Onaga, H. I. & Ogbalu, C. (2008). *The Role of Business in Alleviating Poverty – The Poor as a Source of Innovation and Wealth Creation.* New-Delhi: WCFCG.

Osmani, S. R. (2003). *Evolving Views on Poverty: Concept, Assessment and Strategy.* Manila: Asian Development Bank.

Peredo, A. M. & McLean, M. (2005). Social Entrepreneurship: A Critical Review of the Concept. *Journal of World Business,* pp. 1-29.

Pfeffermann, G. (2001). Poverty Reduction in Developing Countries: The Role of Private Enterprise. *Finance and Development,* 38 (2), pp. 42-45.

Pitta, D. A., Guesalaga, R. & Marshall, P. (2008). The quest for the fortune at the bottom of the pyramid: Potential and challenges. *Journal of Consumer Marketing,* 25 (7), pp. 393-401.

Pless, N. & Maak, T. (2009). Responsible Leaders as Agents of World Benefit: Learning from 'Project Ulysses'. *Journal of Business Ethics,* 85, pp. 59-71.

Prahalad, C. K. & Hammond, A. (2002). Serving the World's Poor Profitably. *Harvard Business Review,* pp. 4-11.

Prahalad, C. K. & Hart, S. L. (2002). The fortune at the bottom of the pyramid. *Strategy and Business,* 26, pp. 44-67.

Prieto-Carron, M., Lund-Thomsen, P., Chan, A., Muro, A. & Bhushan, C. (2006). Critical perspectives on CSR and development: What we know, what we don't know, and what we need to know. *International Affairs,* 82 (5), pp. 977-987.

Raufflet, E. Berranger, A. & Gouin, J.-F. (2008). Innovation in business-community partnerships: Evaluating the impact of local enterprise and global investment models on poverty, bio-diversity and development. *Corporate Governance,* 8 (4), pp. 546-556.

Raufflet, E., Berranger, A. & Aguilar-Platas, A. (2008). Innovative Business Approaches and Poverty: Towards a First Evaluation. In Wankel, C. (Eds.). *Alleviating Poverty through Business Strategy: Global Case Studies in Social Entrepreneurship.* New York: Palgrave Macmillan.

Rein, M. & Stott, L. (2009). Working Together: Critical Perspectives on Six Cross-Sector Partnerships in Southern Africa. *Journal of Business Ethics,* 90, pp. 79-89.

Roberts, R. W. (1992). Determinants of Corporate Social Responsibility Disclosure: An Application of Stakeholder Theory. *Accounting, Organizations and Society,* 17 (6), pp. 595-612.

Rowe, W.-A., Gavrilova, M., Velev, P. & Shaw, T. (2009). Cutting the economic costs of prejudice: Microfinance services in Roma communities in Bulgaria. *Enterprise Development and Microfinance,* 20 (3), pp. 235-246.

Sachs, J. (2008). *Common Wealth: Economics for a Crowded Planet.* London: Penguin Books.

Sasse, C. M. & Trahan, R. T. (2006). Rethinking the new corporate philanthropy. *Business Horizons,* 50, pp. 29-38.

Schwittay, A. (2011). The Marketization of Poverty. *Current Anthropology,* 52 (3), pp. 71-82.

Seelos, C. & Mair, J. (2004). Entrepreneurs in Service of the Poor – Models for Business Contributions to Sustainable Development. *IESE Business School,* pp. 1-7.

Seelos, C. & Mair, J. (2005). Social Entrepreneurship: Creating New Business Models to Serve the Poor. *Business Horizons,* 48, pp. 241-246.

Seelos, C. & Mair, J. (2006). Profitable Business Models and Market Creation in the Context of Deep Poverty: A Strategic View. *IESE Business School,* pp. 1-14.

Selsky, J. & Parker, B. (2005). Cross-sector Partnerships to Address Social Issues: Challenges to Theory and Practice. *Journal of Management,* 12, pp. 849-873.

Sharma, S. (2007). People vs. Poverty: Powering through Partnership. *Futures,* 39, pp. 625-631.

Sharp, J. (2006). Corporate social responsibility and development: An anthropological perspective. *Development Southern Africa,* 23 (2), pp. 213-222.

Shetty, S. (2010). Microcredit, Poverty, and Empowerment: Exploring the Connections. *Perspectives on Global Development and Technology,* 9, pp. 356-391.

Simola, S. K. (2007). The Pragmatics of Care in Sustainable Global Enterprise. *Journal of Business Ethics,* 74, pp. 131-147.

Singer, A. E. (2006). Business Strategy and Poverty Allevition. *Journal of Business Ethics,* 66, pp. 225-231.

Smith, B. R. & Feldman Barr, T. (2007). Reducing Poverty Through Social Entrepreneurship: The Case of Edun. In Stoner, J. A. F. & Wankel, C. (2007). *Innovative Approaches to Reducing Global Poverty.* Charlotte: Information Age Publishing.

Solomon, M. R., Marshall, G. W. & Stuart, E. W. (2008). *Marketing: Real People, Real Choices.* 5th edition. New Jersey: Pearson Education.

Spielman, D. J., Hartwich, F. & von Grebmer, K. (2007). Public-Private Partnerships in International Agricultural Research, *International Food Policy Research Institute,* pp. 1-6.

Sridharan, S. & Viswanathan, M. (2008). Marketing in subsistence marketplaces: consumption and entrepreneurship in a South Indian context. *Journal of Consumer Marketing,* 25 (7), pp. 455-462.

Stefanovic, M. (2007). New Business-NGO Partnerships Help the World's Poorest. *International Trade Forum,* 2, pp. 6-8.

Tang, L. & Li, H. (2009). Corporate social responsibility communication in Chinese and global corporations in China. *Public Relations Review,* 35, pp. 199-212.

Thavisin, N. (2000). A conceptual framework for public private partnership and its applicability to local programmes for social progress in the Asian context. *Regional Development Dialogue,* 21 (2), pp. 98-106.

Thurman, E. (2006). Performance Philanthropy: Bringing Accountability to Charitable Giving. *Harvard International Review,* 1, pp. 18-20.

United Nations Development Programme (2008). *Creating Value for All: Strategies for Doing Business with the Poor.* New York: UNDP.

United Nations Development Programme's Commission on the Private Sector and Development (2004). *Unleashing Entrepreneurship: Making Business Work for the Poor.* New York: UNDP.

van den Waeyenberg, S. (2011). *Essays on the multinational's market introduction of product innovation in less developed countries.* Brussels: Vrije Universiteit Brussel (unpublished).

van Slyke, D. M. & Newman, H. K. (2006). Venture Philanthropy and Social Entrepreneurship in Community Development. *Nonprofit Management and Leadership,* 16 (3), pp. 345-368.

van Tulder, R. (2010). *Transnational Corporations and Poverty Reduction: Strategic and Regional Variations.* London: UNRISD.

van Tulder, R. & da Rosa, A. (2010). *The role of small enterprises in the inclusive business strategies of multinationals.* Rotterdam: The Partnership Resource Center:

van Tulder, R., Fortanier, F. & da Rosa, A. (2011). *Linking Inclusive Business Models and Inclusive Growth.* Rotterdam: The Partnership Resource Center.

van Tulder, R. & van der Zwart, A. (2006). *International Business-Society Management: Linking Corporate Responsibility and Globalization.* New York: Routledge.

Vermeulen, S. & Mayers, J. (2008). Rural poverty reduction through business partnerships? Examples of experiences from the forestry sector. *Environmental Development Sustainability,* 10, pp. 1-18.

Viswanathan, M., Seth, A., Gau, R. & Chaturvedi, A. (2009). Ingraining Product-Relevant Social Good into Business Processes in Subsistence Marketplaces: The Sustainable Market Orientation. *Journal of Macromarketing,* 29 (4), pp. 406-425.

Viswanathan, M., Sridharan, S. & Richtie, R. (2008). Marketing in Subsidence Marketplaces. In Wankel, C. (Eds.). *Alleviating Poverty through Business Strategy: Global Case Studies in Social Entrepreneurship.* New York: Palgrave Macmillan.

von Maltzahn, R. & Durrheim, K. (2008). Is poverty multidimensional? A comparison of income and asset based measures in five Southern African countries. *Social Indicators Research,* 86 (1), pp. 149-162.

Vurro, C., Dacin, M. T. & Perrini, F. (2010). Institutional Antecedents of Partnering for Social Change: How Institutional Logics Shape Cross-Sector Social Partnerships. *Journal of Business Ethics,* 94, pp. 39-53.

Wadham, H. (2009). *Talking about boundaries: Business and NGO perspectives on sustainable development and partnership.* Manchester: Manchester Metropolitan University Business School.

Walsh, J. P. (2005). Book review essay: Taking stock of stakeholder management. *Academy of Management Review,* 30 (2), pp. 426-452.

Walsh, J. P., Kress, J. C. & Beyerchen, K. W. (2005). Book Review Essay: Promises and Perils at the Bottom of the Pyramid. *Administrative Science Quarterly,* pp. 473-482.

Wanderley, L. S. O., Lucian, R., Farache, F. & de Sousa Filho, J. M. (2008). CSR Information Disclosure on the Web: A Context-Based Approach Analysing the Influence of Country of Origin and Industry Sector. *Journal of Business Ethics,* 82, pp. 369-378.

Warhust, A. (2005). Future roles of business in society: The expanding boundaries of corporate responsibility and a compelling case for partnership. *Futures,* 37, pp. 151-168.

Wassmer, U. (2008). Alliance Portfolios: A Review and Research Agenda. *Journal of Management,* 36, pp. 141-171.

WECD (1987). *Our common future.* UK: Oxford University Press.

Weerawardena, J. & Mort, G. S. (2006). Investigating social entrepreneurship: A multidimensional model. *Journal of World Business,* 41, pp. 21-35.

Williams, C. A. & Aguilera, R. V. (2008). *Corporate Social Responsibility in a Comparative Perspective.* Champaign: University of Illinois at Urbana-Champaign.

World Bank (2008). *2008 World Development Indicators – Poverty Data – A Supplement to World Development Indicators 2008.* Washington: World Bank.

World Bank (2011). *What is Poverty and Why to Measure It?* Washington: Word Bank.

World Business Council for Sustainable Development (2004). *Doing business with the poor: A field guide.* Switzerland: Atlar Toto Presse SA.

World Business Council for Sustainable Development (2006). *From Challenge to Opportunity – The role of business in tomorrow's society.* Switzerland: WBCSD.

Wright, G. A. N. & Dondo, A. (2001). *"Are You Poor Enough?" – Client Selection by Microfinance Institutions.* Nairobi: MicroSave.

Yunus, M. (2006). *Social Business Entrepreneurs are the Solution.* Boston: Skoll World Forum on Social Entrepreneurship.

Yunus, M. (2008). *Die Armut besiegen.* Carl Hanser Verlag: München.

Yunus, M. (2010). *Building Social Business: The New Kind of Capitalism that Serves Humanity's Most Pressing Needs.* New York: Public Affairs.

Yunus, M. & Joilis, A. (1999). *Banker to the Poor: Micro-Lending and the Battle Against World Poverty.* New York: Persus Books Group.

Yunus, M., Moingeon, B. & Lehmannn-Ortega, L. (2010). Building Social Business Models: Lessons from the Grameen Experience. *Longe Range Planning,* 43, pp. 308-325.

Zahra, S. A., Gedajlovic, E., Neubaum, D. O. & Shulman, J. M. (2009). A typology of social entrepreneurs: Motives, search processses and ethical challenges. *Journal of Business Venturing,* 24, pp. 519-532.

General Websites

Fortune website (2011) [online]. Available at: http://money.cnn.com/magazines/fortune/global500/2011/countries/US.html. [Accessed 15 October 2011].

Industry Qualification Benchmark website (2011) [online]. Available at: http://www.nyse.com/indexes/nyaindex.csv [Accessed 16 October 2011].

Tagesschau website (2011) [online]. *Wegen Ölverschmutzung im Regenwald: Gericht in Equador verurteilt Chevron zu Millardenstrafe.*February 15, 2011. Available at: http://www.tagesschau.de/wirtschaft/chevron102.html [Accessed 18 November 2011].UNDP website (2011) [online]. Available at: http://hdr.undp.org/en/statistics/mpi/ [Accessed 18 November 2011].

Company Websites

Allianz website [online]. Available at: www.allianz.com. [Accessed 23 October 2011].

American International Group [online]. Available at: www.aigcorporate.com. [Accessed 23 October 2011].

Amerisource Bergen [online]. Available at: www.amerisourcebergen.com. [Accessed 23 October 2011].

Arcelor Mittal website [online]. Available at: www.arcelormittal.com. [Accessed 23 October 2011].

Assicurazioni Generalli website [online]. Available at: www.generali.com. [Accessed 23October 2011].

AT & T website [online]. Available at: www.att.com. [Accessed 23 October 2011].

Aviva website [online]. Available at: www.aviva.com. [Accessed 23 October 2011].

AXA website [online]. Available at: www.axa.com. [Accessed 23 October 2011].

Banco Santander website [online]. Available at: www.santander.com. [Accessed 23 October 2011].

Bank of America Corporation website [online]. Available at: www.bankofamerica.com. [Accessed 24 October 2011].

BASF website [online]. Available at: www.basf.com. [Accessed 24 October 2011].

Berkshire Hathaway website [online]. Available at: www.berkshirehathaway.com. [Accessed 24 October 2011].

BMW website [online]. Available at: www.bmwgroup.com. [Accessed 24 October 2011].

BNP Paribas website [online]. Available at: www.bnpparibas.com. [Accessed 24 October 2011].

BP website [online]. Available at: www.bp.com. [Accessed 24 October 2011].

Cardinal Health website [online]. Available at: www.cardinal.com. [Accessed 24 October 2011].

Carrefour website [online]. Available at: www.carrefour.com. [Accessed 24 October 2011].

Chevron website [online]. Available at: www.chevron.com. [Accessed 24 October 2011].

China Mobile Communications website [online]. Available at: www.chinamobile.com. [Accessed 25 October 2011].

China National Petroleum website [online]. Available at: www.cnpc.com.cn. [Accessed 25 October 2011].

China Railway Group website [online]. Available at: www.crec.cn. [Accessed 25 October 2011].

Citigroup website [online]. Available at: www.citigroup.com. [Accessed 25 October 2011].

ConocoPhillips website [online]. Available at: www.conocophillips.com. [Accessed 25 October 2011].

Costco Wholesale website [online]. Available at: www.costco.com. [Accessed 25 October 2011].

Crédit Agricoles website [online]. Available at: www.credit-agricole.com. [Accessed 25 October 2011].

CVS Caremark website [online]. Available at: www.cvscaremark.com. [Accessed 25 October 2011].

Daimler website [online]. Available at: www.daimler.com/dccom/home/de. [Accessed 25 October 2011].

Deutsche Post website [online]. Available at: www.dp-dhl.com. [Accessed 26 October 2011].

Deutsche Telekom website [online]. Available at: www.telekom.com. [Accessed 26 October 2011].

Dexia Group website [online]. Available at: www.dexia.com. [Accessed 26 October 2011].

E.On website [online]. Available at: www.eon.com. [Accessed 26 October 2011].

Électricité de France website [online]. Available at: www.edf.fr. [Accessed 26 October r 2011].

Enel website [online]. Available at: www.enel.com. [Accessed 26 October 2011].

ENI website [online]. Available at: www.eni.it. [Accessed 26 October 2011].

EXOR Group [online]. Available at: www.exor.com. [Accessed 26 October 2011].

Exxon Mobil website [online]. Available at: www.exxonmobil.com. [Accessed 26 October 2011].

Fannie Mae website [online]. Available at: www.fanniemae.com. [Accessed 27 October 2011].

Ford Motor website [online]. Available at: www.ford.com. [Accessed 27 October 2011].

Freddie Mac website [online]. Available at: www.freddiemac.com. [Accessed 27 October 2011].

Gazprom website [online]. Available at: www.gazprom.com. [Accessed 27 October 2011].

GDF Suez website [online]. Available at: www.gdfsuez.com. [Accessed 27 October 2011].

General Electrics website [online]. Available at: www.ge.com. [Accessed 27 October 2011].
General Motors website [online]. Available at: www.gm.com. [Accessed 27 October 2011].
Glencore International website [online]. Available at: www.glencore.com. [Accessed 27 October 2011].
Groupe BPCE website [online]. Available at: www.bpce.fr. [Accessed 27 October 2011].
Hewlett-Packard website [online]. Available at: www.hp.com. [Accessed 27 October 2011].
Hitachi website [online]. Available at: www.hitachi.com. [Accessed 28 October 2011].
Hon Hai Precision Industry website [online]. Available at: www.foxconn.com. [Accessed 28 October 2011].
Honda Motor website [online]. Available at: www.honda.com. [Accessed 28 October 2011].
HSBC Holdings website [online]. Available at: www.hsbc.com. [Accessed 28 October 2011].
Hyundai Motor website [online]. Available at: www.hyundai.com. [Accessed 28 October 2011].
Indian Oil website [online]. Available at: www.iocl.com. [Accessed 28 October 2011].
Industrial and Commercial Bank of China website [online]. Available at: www.icbc.com.cn. [Accessed 28 October 2011].
ING Group website [online]. Available at: www.ing.com. [Accessed 28 October 2011].
International Business Machines website [online]. Available at: www.ibm.com. [Accessed 29 October 2011].
J. P. Morgan Chase & Co. website [online]. Available at: www.jpmorganchase.com. [Accessed 29 October 2011].
Japan Post Holdings website [online]. Available at: www.japanpost.jp. [Accessed 29 October 2011].
JX Holdings website [online]. Available at: www.hd.jx-group.co.jp. [Accessed 29 October 2011].
Kroger website [online]. Available at: www.kroger.com. [Accessed 29 October 2011].
Llyods Banking Group website [online]. Available at: www.lloydsbankinggroup.com. [Accessed 29 October 2011].
Lukoil website [online]. Available at: www.lukoil.com. [Accessed 29 October 2011].
Marathon Oil website [online]. Available at: www.marathonoil.com. [Accessed 29 October 2011].
McKesson website [online]. Available at: www.mckesson.com. [Accessed 30 October 2011].
Metro Group website [online]. Available at: www.metrogroup.de. [Accessed 30 October 2011].
Munich Re Group website [online]. Available at: www.munichre.com. [Accessed 30 October 2011].
Nestlé website [online]. Available at: www.nestle.com. [Accessed 30 October 2011].
Nippon Life Insurance website [online]. Available at: www.nissay.co.jp. [Accessed 30 October 2011].
Nippon Telegraph & Telephone website [online]. Available at: www.ntt.co.jp. [Accessed 30 October 2011].
Nissan Motors website [online]. Available at: www.nissan-global.com. [Accessed 30 October 2011].

Panasonic website [online]. Available at: www.panasonic.net. [Accessed 30 October 2011].

PDVSA website [online]. Available at: www.pdvsa.com. [Accessed 30 October 2011].

Pemex website [online]. Available at: www.pemex.com. [Accessed 30 October 2011].

Petrobras website [online]. Available at: www.petrobras.com.br. [Accessed 30 October 2011].

Petronas website [online]. Available at: www.petronas.com.my. [Accessed 31 October 2011].

Peugeout website [online]. Available at: www.psa-peugeot-citroen.com. [Accessed 31 October 2011].

Procter & Gamble website [online]. Available at: www.pg.com. [Accessed 31 October 2011].

Prudential website [online]. Available at: www.prudential.co.uk. [Accessed 31 October 2011].

Repsol YPF website [online]. Available at: www.repsol.com. [Accessed 31 October 2011].

Royal Bank of Scotland website [online]. Available at: www.rbs.com. [Accessed 31 October 2011].

Royal Dutch Shell website [online]. Available at: www.shell.com. [Accessed 31 October 2011].

Samsung Electronics website [online]. Available at: www.samsung.de. [Accessed 31 October 2011].

Siemens website [online]. Available at: www.siemens.com. [Accessed 31 October 2011].

Sinopec Group website [online]. Available at: www.sinopec.com. Accessed 1 November 2011].

SK Holdings website [online]. Available at: www.sk.com. [Accessed 1 November 2011].

Société Générale website [online]. Available at: www.societegeneral.com. [Accessed 1 November 2011].

Sony website [online]. Available at: www.sony.net. [Accessed 1 November 2011].

State Grid website [online]. Available at: www.sgcc.com.cn. [accessed 1 November 2011].

Statoil website [online]. Available at: www.statoil.com. [Accessed 1 November 2011].

Telefónica website [online]. Available at: www.telefonica.com. [Accessed 1 November 2011].

Tesco website [online]. Available at: www.tescoplc.com. [Accessed 1 November 2011].

Toshiba website [online]. Available at: www.toshiba.co.jp. [Accessed 1 November 2011].

Total website [online]. Available at: www.total.com. [Accessed 2 November 2011].

Toyota Motor website [online]. Available at: www.toyota-global.com. [Accessed 2 November 2011].

United Health Group website [online]. Available at: www.unitedhealthgroup.com. [Accessed 2 November 2011].

Valero Energy website [online]. Available at: www.valero.com. [Accessed 2 November 2011].

Verizon Communications website [online]. Available at: www.verizon.com. [Accessed 2 November 2011].

Vodafone website [online]. Available at: www.vodafone.com. [Accessed 2 November 2011].

Volkswagen website [online]. Available at: www.volkswagenag.com. [Accessed 2 November 2011].

Wal-Mart Stores website [online]. Available at: www.walmartstores.com. [Accessed 2 November 2011].

Wells Fargo website [online]. Available at: www.wellsfargo.com. [Accessed 2 November 2011].

Company Reports

Allianz Group (2011). *Sustainable Development Report 2010/11.*

Arcelor Mittal (2010). *Safe Sustainable Steel – Corporate Responsibility Report 2010.*

Assicurazioni Generalli (2010). *Sustainability Report 2010.*

AT&T (2010). *Meet the Possibility Economy – 2010 AT&T Sustainability Report.*

Aviva (2010). *Corporate Responsibility Report 2010.*

AXA (2010). *Activity and Corporate Responsibility Report 2010.*

Banco Santander (2010). *Sustainability Report 2010.*

Bank of America Corporation (2010). *Opportunity in Motion – Corporate Social Responsibility Report 2010.*

BASF (2010). *BASF Report 2010 – Economic, Environmental and Social Performance.*

BMW (2010). *Sustainable Value Report 2010.*

BNP Paribas (2010). *Corporate Social Responsibility Report 2010.*

BP (2010). *Sustainability Review 2010.*

Cardinal Health (2010). *Giving with Purpose – Community Report 2010.*

Carrefour (2010). *Reinventing Carrefour – Annual Activity and Sustainability Report 2010.*

Chevron (2010). Corporate Responsibility Report 2010.

China Mobile Communications (2010). *China Mobile: Enabling a Better Life – Sustainable Future with ICT – China Mobile Limited 2010 Sustainability Report.*

China National Petroleum (2010). *Corporate Social Responsibility Report 2010 – Caring for Energy - Caring for You.*

China Railway Group (2009). *Social Responsibility Report of 2009.*

Citigroup (2010). *Global Citizenship Report 2010.*

ConocoPhillips (2009). *2009 Sustainable Development Review.*

Costco Wholesale (2009). *Corporate Sustainability Report.*

Crédit Agricoles (2010). *The 2010 Sustainable Development Compendium.*

CVS Caremark (2010). *Innovation Leadership Commitment – 2010 Corporate Social Responsibility Report.*

Daimler (2010). *360 Facts on Sustainability 2010.*

Deutsche Post (2010). *Living Responsibility – Corporate Responsibility Report 2010.*

Deutsche Telekom (2010). *The 2010 Corporate Responsibility Report. We Take Responsibility.*

Dexia Group (2010). *Dexia Sustainable Development Report 2010.*

E.ON (2010). *E.ON CR Report 2010.*

Électricité de France (2010). *EDF Group Activity & Sustainable Development 2010.*

Enel (2010). *Sustainability Report 2010.*

ENI (2010). *ENI for Development 2010.*

ENI (2010). *Sustainability Performance 2010.*

Exxon Mobil (2010). *2010 Corporate Citizenship Report.*

Ford Motor (2010). *Sustainability Report 2009/10.*

Gazprom (2010). *OAO Gazprom Annual Report 2010.*

Gazprom (2010). *OAO Gazprom Environmental Report 2010.*

GDF Suez (2010). *Sustainable Development Report 2010.*

General Electrics (2010). *Sustainable Growth – GE 2010 Citizenship Report.*

General Motors. *Corporate Social Responsibility Report 2009.*

Glencore International (2010). *Glencore Sustainability Report 2010.*

Groupe BCPE (2010). *Review of Operations and Sustainable Development Report 2010.*

Hewlett-Packard (2010). *A Connected World – The Impact of HP Global Citizenship 2010 – and beyond.*

Hitachi (2010). *Hitachi Group. Corporate Sustainability Report 2010.*

Hon Hai Precision Industry (2010). *Foxconn 2010 Corporate Social and Environmental Responsibility Annual Report.*

Honda Motor (2011). *Striving to Be a Company Society Wants to Exist – CSR Report 2011.*

HSBC Holdings (2010). *Community Investment at HBSC 2010.*

HSBC Holdings (2010). *Sustainability Report 2010.*

Hyundai Motor (2011). *The Road To Sustainability – Hyundai Motor Company 2011 Sustainability Report.*

Indian Oil (2010). *Enduring Offerings – Sustainability Report 2009-10.*

Industrial and Commercial Bank of China (2010). *Corporate Social Responsibility Report 2010.*

ING Group (2010). *Corporate Responsibility Report 2010 – ING in Society.*

International Business Machines (2010). *2010 Corporate Responsibility Summary.*

J. P. Morgan Chase & Co. (2010). *The Way Forward – 2010 Corporate Responsibility Report.*

Japan Post Holdings (2010). *Japan Post Group Annual Report 2010.*

JX Holdings (2010). *JX Report for a Sustainable Future 2010 – The Future of Energy, Resources and Materials.*

Kroger (2010). *Doing Our part – 2010 Sustainability Report.*

Llyods Banking Group (2010). *Responsible Business Report 2010.*

Lukoil (2010). *2009-2010 Sustainability Report.*

Marathon Oil (2010). *Living Our Values – 2010 Corporate Social Responsibility Report.*

McKesson (2010). *Corporate Citizenship 2010.*

Metro Group (2010). *Sustainability- Progress Report – Key Data and Targets.*

Munich Re Group (2010). *Munich RE Corporate Responsibility 2010.*

Nestlé (2010). *Nestlé Creating Shared Value and Rural Development Report 2010.*

Nippon Telegraph & Telephone (2010). *NTT Group CSR Report 2010.*

Nissan Motors (2010). *Nissan: Enriching People's Lives – Sustainability Report 2010.*

Panasonic (2011). *Sustainability Report 2011.*

Pemex (2010). *Pemex Group Social Responsibility 2010.*

Petrobras (2009). *Sustainability Report.*

Petronas (2010). *2010 Sustainability Report.*

Peugeout (2010). *2010 Sustainable Development Performance Indicators.*

Procter & Gamble (2011). *2011 Sustainability Report – Commitment to Everyday Life.*

Prudential (2010). *Corporate Responsibility 2010 –Long-term Thinking.*

Repsol YPF (2010). *Corporate Responsibility Report 2010 – Our Relationships.*

Royal Bank of Scotland (2010). *Building a Sustainable RBS – Sustainability Report 2010.*

Royal Dutch Shell (2010). *Royal Dutch Shell PLC Sustainability Report 2010.*

Samsung Electronics (2010). *2009-2010 Sustainability Report.*

Siemens (2010). *Sustainability Report 2010 – Seize Opportunities, Minimize Risks, Live Our Values.*

Sinopec Group (2010). *Making Every Drop Count – 2010 Corporate Social responsibility Report.*

SK Holdings (2009). *We Deliver Happiness.*

Société Générale (2011). *Corporate Social Responsibility at Societe Generale 2011.*

Sony (2010). *CSR Report 2010.*

State Grid (2009). *2009 Corporate Social responsibility Report of State Grid Corporation of China.*

Statoil (2010). *Annual Report 2010.*

Telefónica (2010). *Annual Corporate Responsibility and Sustainability Report Telefonica S. A.*

Tesco (2011). *Corporate Responsibility Report 2011.*

Toshiba (2011). *CSR Report – Corporate Social Responsibility 2011.*

Total (2010). *Society and Environment Report 2010.*

Toyota Motor (2011). *Sustainability Report 2011.*

United Health Group (2010). *Our Commitment to Healthier Communities – 2010 Social Responsibility Report.*

Valero Energy (2010). *2010 Social Responsibility Report.*

Verizon Communications (2011). *Verizon Communications Corporate Responsibility Report 2010/11.*

Vodafone (2010). *Sustainability Report 2011.*

Volkswagen (2010). *Driving Ideas – Sustainability Report 2009/2010.*

Wal-Mart Stores (2011). *Building the Next Generation Walmart Responsibly – 2011 Global Responsibility Report.*

Wells Fargo (2010). *Standing Together with Our Communities – Corporate Social responsibility Interim Report 2010.*

EINZELSCHRIFTEN

Marc Jizba
Die Nachhaltigkeitsleistung der deutschen DAX30-Unternehmen – Eine kritische Bewertung des Status Quo mit Hilfe des Sustainable-Value-Added-Konzepts
Lohmar – Köln 2014 • 172 S. • € 48,- (D) • ISBN 978-3-8441-0366-3

Antonio Vera
Spekulationsblasen in der frühen Neuzeit – Ein systematischer Vergleich der Ursachen, Mechanismen und Auswirkungen der Mississippi und der South Sea Bubble
Lohmar – Köln 2015 • 168 S. • € 47,- (D) • ISBN 978-3-8441-0379-3

Georg Baltes
New Perspectives on Supply and Distribution Chain Financing: Case Studies from China and Europe
Lohmar – Köln 2015 • 396 S. • € 66,- (D) • ISBN 978-3-8441-0384-7

Katja Müller
Wertschaffende Kooperationsbeziehungen – Kooperationsbeziehungen im Lean Management analysiert aus einer konstruktivistischen Sicht
Lohmar – Köln 2015 • 364 S. • € 64,- (D) • ISBN 978-3-8441-0385-4

Isabel Arnold
Personalentwicklung von Führungskräften in Zeiten von Change – Eine Betrachtung aus Sicht des systemorientierten Managements
Lohmar – Köln 2015 • 236 S. • € 56,- (D) • ISBN 978-3-8441-0389-2

Ansgar Kernder
Bewertung der Emittentenqualität am Markt für Mittelstandsanleihen – Eine empirische Analyse im Lichte der Prinzipal-Agenten-Theorie
Lohmar – Köln 2015 • 436 S. • € 68,- (D) • ISBN 978-3-8441-0392-2

Kathrin Bischoff
Business Approaches to Poverty Alleviation – A Classification of Company Initiatives
Lohmar – Köln 2014 • 208 S. • € 49,- (D) • ISBN 978-3-8441-0393-9